Preparing
NEPA
Environmental
Assessments

A User's Guide to Best Professional Practices

Charles Eccleston
J. Peyton Doub

Preparing NEPA Environmental Assessments

A User's Guide to Best Professional Practices

CRC Press
Taylor & Francis Group
Boca Raton   London   New York

CRC Press is an imprint of the
Taylor & Francis Group, an **informa** business

CRC Press
Taylor & Francis Group
6000 Broken Sound Parkway NW, Suite 300
Boca Raton, FL 33487-2742

First issued in paperback 2017

© 2012 by Taylor & Francis Group, LLC
CRC Press is an imprint of Taylor & Francis Group, an Informa business

No claim to original U.S. Government works

Version Date: 20120409

ISBN 13: 978-1-4398-0882-5 (hbk)
ISBN 13: 978-1-138-07505-4 (pbk)

This book contains information obtained from authentic and highly regarded sources. Reasonable efforts have been made to publish reliable data and information, but the author and publisher cannot assume responsibility for the validity of all materials or the consequences of their use. The authors and publishers have attempted to trace the copyright holders of all material reproduced in this publication and apologize to copyright holders if permission to publish in this form has not been obtained. If any copyright material has not been acknowledged please write and let us know so we may rectify in any future reprint.

Except as permitted under U.S. Copyright Law, no part of this book may be reprinted, reproduced, transmitted, or utilized in any form by any electronic, mechanical, or other means, now known or hereafter invented, including photocopying, microfilming, and recording, or in any information storage or retrieval system, without written permission from the publishers.

For permission to photocopy or use material electronically from this work, please access www.copyright.com (http://www.copyright.com/) or contact the Copyright Clearance Center, Inc. (CCC), 222 Rosewood Drive, Danvers, MA 01923, 978-750-8400. CCC is a not-for-profit organization that provides licenses and registration for a variety of users. For organizations that have been granted a photocopy license by the CCC, a separate system of payment has been arranged.

Trademark Notice: Product or corporate names may be trademarks or registered trademarks, and are used only for identification and explanation without intent to infringe.

Visit the Taylor & Francis Web site at
http://www.taylorandfrancis.com

and the CRC Press Web site at
http://www.crcpress.com

Contents

Preface xiii
Introduction xv

Section I NEPA environmental assessment process

Chapter 1 Overview and historical development of NEPA 3
1.1 Introduction 3
1.2 Origin of environmental movement 3
    1.2.1 Nineteenth century 4
    1.2.2 Early twentieth century 5
    1.2.3 Environmental decade of the 1960s 5
1.3 Historical development of NEPA 6
    1.3.1 Birth of NEPA 6
    1.3.2 "Detailed statement" 8
    1.3.3 Passage of NEPA 9
    1.3.4 NEPA's implementation 10
    1.3.5 NEPA's implementation regulations 11
    1.3.6 Public comment review process 12
    1.3.7 NEPA today 12
    1.3.8 NEPA's effect around the world 14
    1.3.9 NEPA's global precedent 14
1.4 Brief overview of NEPA 15
    1.4.1 Procedure for passing laws and regulations 15
    1.4.2 Regulatory nomenclature 16
    1.4.3 Overview of NEPA's mandate 16
    1.4.4 Purpose of NEPA 17
    1.4.5 Title I—declaration of the National Environmental
          Policy Act 17
    1.4.6 Title II—Council on Environmental Quality 19
1.5 NEPA regulations 20
1.6 Overview of the NEPA planning process 21
    1.6.1 When must NEPA begin? 21

	1.6.2	When must NEPA be completed?...22
	1.6.3	Three levels of NEPA compliance 22
	1.6.4	Initiating the NEPA process..22
	1.6.5	Environmental assessment..25
	1.6.6	Environmental impact statement 27

References...27

Chapter 2 General concepts and requirements 29
2.1 Introduction..29
2.2 NEPA is a planning and decision making process.......................30
    2.2.1 Reasonable alternatives..30
2.3 Interim actions 30
    2.3.1 Eligibility for interim action status 31
2.4 Integrating NEPA with other requirements 32
    2.4.1 Integrating environmental design arts............................32
2.5 Conducting an early and open process 32
2.6 Public involvement 33
2.7 Scoping..34
2.8 Systematic and interdisciplinary planning 34
    2.8.1 Systematic 34
    2.8.2 Interdisciplinary 34
2.9 Writing documents in plain English..35
2.10 Incorporation by reference 35
2.11 Adopting another agency's EA 36
2.12 Methodology 36
2.13 Fair and objective analysis 37
2.14 Dealing with incomplete and unavailable information..................37
    2.14.1 Incomplete information 38
    2.14.2 Unavailable information..38
References...39

Chapter 3 NEPA and environmental impact analysis..........................41
3.1 Introduction..41
3.2 Actions...42
3.3 Environmental disturbances..44
3.4 Receptors and resources 44
3.5 Impact analysis 45
3.6 Significance...45
3.7 Mitigation and monitoring...46
    3.7.1 Mitigation...46
    3.7.2 Monitoring 47
References...48

Chapter 4 Threshold question: Determining whether an EA or an EIS is required 49
4.1 Introduction 49
4.2 Proposals 50
4.3 Legislation 50
4.4 Major 51
4.5 Federal 51
4.6 Actions 52
4.7 Significance 52
    4.7.1 Context 53
    4.7.2 Intensity 54
4.8 Affecting 54
4.9 Human environment 55
4.10 Categorical exclusions 56
    4.10.1 Guidance on applying categorical exclusions 57
    4.10.2 Documenting a categorical exclusion 57
    4.10.3 Public engagement and disclosure 58
4.11 Long versus short environmental assessments 58
References 60

Chapter 5 Environmental assessment process 61
5.1 Introduction 61
5.2 Planning the environmental assessment 61
    5.2.1 Public involvement 63
    5.2.2 Identifying alternatives 64
    5.2.3 Determining the potential for significance 65
    5.2.4 Three purposes that an EA may serve 66
    5.2.5 Timing 66
    5.2.6 EAs are public documents 67
        5.2.6.1 Public notification 67
        5.2.6.2 Consultation 68
        5.2.6.3 Scoping and public meetings 69
    5.2.7 Applicants and environmental assessment contractors 69
5.3 Analysis in an EA 70
    5.3.1 Decision-based scoping 70
    5.3.2 Analysis of impacts 71
    5.3.3 Attention centered on proposed action 71
    5.3.4 Significance determinations are reserved for the FONSI .. 72
5.4 Streamlining the EA compliance process 72
    5.4.1 Reducing the length of an EA 73
    5.4.2 Cooperating with other agencies 75
    5.4.3 Is time money? 75
    5.4.4 Tiering 76
5.5 Issuing a finding of no significant impact 77

	5.5.1	FONSI is a public document	77
	5.5.2	Waiting period	77
5.6	Administrative record		78
	5.6.1	Administrative record and case law	78
5.7	Serving as an expert witness		79
	5.7.1	Are you an expert witness?	79
	5.7.2	Discovery	80
	5.7.3	Depositions and reports	80
	5.7.4	Trial	80
References			81

Chapter 6 Environmental impact assessment 83

- 6.1 Introduction 83
- 6.2 Sliding-scale approach 83
  - 6.2.1 Recommendations 84
- 6.3 Identifying alternatives 85
  - 6.3.1 Types of alternatives 86
  - 6.3.2 Range of reasonable alternatives 86
    - 6.3.2.1 Appropriate number of alternatives to include in the analysis 88
    - 6.3.2.2 Art of defining a range of reasonable alternatives 88
    - 6.3.2.3 Comparing alternatives 90
    - 6.3.2.4 General recommendations 90
  - 6.3.3 No-action alternative 91
  - 6.3.4 Recommendations for identifying and investigating alternatives 91
- 6.4 Impacts 91
  - 6.4.1 Bounding analysis 93
  - 6.4.2 Analysis of impacts 94
    - 6.4.2.1 Biological impacts 95
  - 6.4.3 Incomplete or unavailable information 96
  - 6.4.4 Cumulative effects analysis in EAs 97
    - 6.4.4.1 Purpose of the analysis 98
    - 6.4.4.2 Performing the analysis 98
  - 6.4.5 General recommendations 103
    - 6.4.5.1 Segmentation 105
- 6.5 Human health effects 106
  - 6.5.1 Recommendations 107
- 6.6 Accident analyses 107
  - 6.6.1 Sliding scale 108
  - 6.6.2 Overview 108
  - 6.6.3 Accident scenarios and probabilities 109
    - 6.6.3.1 Range of accident scenarios 109

		6.6.3.2	Scenario probabilities	110
	6.6.4	Risk		111
		6.6.4.1	Conservatisms	111
	6.6.5	Accident consequences		111
		6.6.5.1	Involved and noninvolved workers	111
		6.6.5.2	Uncertainty	112
	6.6.6	Intentionally destructive acts		112
6.7	Considering other environmental requirements			113
	6.7.1	Clean air conformity requirements		113
		6.7.1.1	Recommendations	114
	6.7.2	Floodplain and wetland environmental review requirements		114
	6.7.3	National Historic Preservation Act (16 U.S.C. 470)		115
	6.7.4	Other related legal requirements		116
		6.7.4.1	Executive order requirements	116
		6.7.4.2	Other statutory requirements	117
6.8	Environmental justice			119
	6.8.1	General guidance		120
		6.8.1.1	CEQ guidance	120
		6.8.1.2	Environmental Protection Agency guidance	121
		6.8.1.3	Department of Energy guidance	121
	6.8.2	Determining the appropriate level of NEPA review		121
		6.8.2.1	Categorical exclusions	121
		6.8.2.2	Environmental assessments	121
	6.8.3	Analyzing environmental impacts		122
	6.8.4	Evaluating high and adverse impacts		122
		6.8.4.1	Factors used in determining whether an impact is disproportionately high and adverse	123
		6.8.4.2	Focusing on impacts that would be different	124
		6.8.4.3	Unique pathways, exposures, and cultural practice	125
		6.8.4.4	Considering cumulative impacts	126
	6.8.5	Assessing significance and mitigation measures		127
	6.8.6	CEQ guidance on mitigation and monitoring		127
		6.8.6.1	Addressing mitigation in environmental assessments	127
		6.8.6.2	Ineffective mitigation measures	128
		6.8.6.3	Environmental monitoring	128
		6.8.6.4	Implementing mitigation measures	129
		6.8.6.5	Transparency in implementing mitigation measures and a monitoring program	130
References				130

Chapter 7		**Writing the environmental assessment**	**133**
7.1	Introduction		133
7.2	General direction for preparing an EA		134
	7.2.1	Proposals and contracting	134
	7.2.2	Staffing	137
	7.2.3	Research	139
	7.2.4	Documenting assumptions	140
	7.2.5	"Will" versus "would"	141
	7.2.6	Readability and plain language	141
		7.2.6.1 Style	142
		7.2.6.2 Plain language	143
	7.2.7	Visuals and graphic aids	143
		7.2.7.1 Figures	144
		7.2.7.2 Tables	144
		7.2.7.3 Photographs and photosimulations	145
		7.2.7.4 Other techniques for improving readability	146
	7.2.8	Glossaries	146
	7.2.9	Other writing style notes	147
		7.2.9.1 Measurements	147
		7.2.9.2 Species names	147
	7.2.10	How long should the EA be?	148
	7.2.11	Editing	148
7.3	Typical contents of an environmental assessment		148
	7.3.1	Suggested EA outline	149
	7.3.2	Organization	150
7.4	Performing the analysis		151
	7.4.1	Five-step methodology	151
7.5	Specific documentation requirements and guidance		157
	7.5.1	Need section	158
	7.5.2	Affected environment section	159
	7.5.3	Proposed action and alternatives section	161
	7.5.4	Environmental impact (environmental consequences) section	165
	7.5.5	Listing permits, approvals, and regulatory requirements	172
	7.5.6	Listing of agencies and people consulted	173
	7.5.7	List of preparers	175
References			176
Chapter 8		**Assessing significance**	**177**
8.1	Introduction		177
8.2	Definitions and use of the term "significance"		177
	8.2.1	Potential significance should determine the depth of analysis	178

Contents   xi

	8.2.2	Need for a systematic approach 179
8.3	General procedure for determining significance 179	
	8.3.1	Description and basis of the tests 180
	8.3.2	Assessing the seven tests of significance 182
	8.3.3	Additional significance considerations in practice 196
	8.3.4	Evidence of significance 196
8.4	Significant beneficial impacts 197	
	8.4.1	*Hiram Clarke Civic Club v. Lynn* 198
	8.4.2	*Douglas County v. Babbitt* 198
	8.4.3	Conclusion 200
References 201		

Chapter 9   Finding of no significant impacts 203
9.1   Introduction 203
9.2   Reaching a determination of nonsignificance 204
       9.2.1   Principles governing sound decision making 205
       9.2.2   Criteria that must be met in reaching a decision to not prepare an EIS 205
       9.2.3   Agency's administrative record 207
9.3   Preparing the FONSI 208
       9.3.1   Documentation requirements 209
       9.3.2   A checklist for preparing the FONSI 211
9.4   Mitigated FONSIs 211
       9.4.1   Mitigation and the courts 213
       9.4.2   Criteria for adopting mitigation measures 215
References 215

Section II NEPA case law and non-NEPA environmental assessment documents

Chapter 10 An overview of NEPA law and litigation 219
10.1 Principles underlying legal interpretations of NEPA 219
10.2 Process for passing laws and regulations 220
       10.2.1 United States Statutes at Large 220
       10.2.2 United States Code 221
       10.2.3 Code of Federal Regulations 222
10.3 Interpreting NEPA's regulatory requirements 223
       10.3.1 What constitutes "reasonable"? 224
10.4 A substantive versus procedural process 225
       10.4.1 NEPA is essentially a procedural process 225
10.5 Requirements for bringing a successful environmental lawsuit .. 227
       10.5.1 Ripeness 227
              10.5.1.1 Criteria for "standing" 228

		10.5.1.2	Mootness	229
10.6	Litigation process			229
	10.6.1	Enforcement of NEPA		229
	10.6.2	Legal system		230
	10.6.3	Bringing a suit against an agency		231
	10.6.4	Legal strategies		232
	10.6.5	Court's role		233
	10.6.6	Legal remedy		234
	10.6.7	Reviewing an agency's administrative record		234
10.7	Legal standards applied in reviewing NEPA			236
	10.7.1	Administrative Procedures Act		236
	10.7.2	Judicial review standards		237
		10.7.2.1	Disagreement among experts	237
	10.7.3	Court's role		238
		10.7.3.1	Determining whether an EIS is required	238
		10.7.3.2	Determining if an EIS needs to be supplemented	240
		10.7.3.3	Agency's final decision	241
		10.7.3.4	What the courts are looking for	242
10.8	Judicial remedies in NEPA suits			243
	10.8.1	Recovering legal costs		244
	10.8.2	Types of remedies		244
10.9	Recent case law involving NEPA			245
10.10	EA case law and EAs			246
10.11	Study of alternatives analysis in NEPA documents			248
10.12	Other cases reflecting on the preparation of EAs			249
	10.12.1	Avoid implying that a FONSI is predetermined		249
	10.12.2	Participate actively as cooperating agency		249
References				250

Chapter 11 Specialized non-NEPA environmental assessment documents 253

11.1	Introduction	253
11.2	Phase I environmental site assessment	253
11.3	Cell tower NEPA review	259
11.4	Biological assessment	262
11.5	Environmental assessments and state NEPA programs	266
References		271

Chapter 12 Summary 273
12.1	Epilogue	273

Index 275

Preface

Since its publication in 2001, *Effective Environmental Assessments: How to Manage and Prepare NEPA Assessments*, has provided practitioners under the National Environmental Policy Act (NEPA) with the best available technical guidance book specifically focused on the environmental assessment (EA) rather than the environmental impact statement (EIS). Most NEPA reference books deal almost exclusively with the EIS and relegate the EA to at best one chapter and often only to a brief mention. They give the impression that the EIS is the primary impact assessment documentation used under NEPA.

Although the EIS is certainly the best known and most publicly visible impact assessment document prepared under NEPA, the EA is increasingly becoming the unsung workhorse of NEPA. While the EIS traces its origin to the original statutory wording of NEPA, the EA was borne of practicality issues that only became apparent once the NEPA process was actually put into practice. Agencies could not meet the environmental objectives of NEPA without considering the potential impacts of their relatively minor actions, and they could not effectively carry out their missions within their allocated budgets if they had to prepare an EIS for each of those actions. Since the Council on Environmental Quality (CEQ) included basic definitions and direction for the EA in its NEPA regulations in 1978, agencies have prepared thousands of EAs each year for actions that might have previously either received no attention under NEPA or been the subject of useless EISs that needlessly wasted taxpayer resources and mission-critical time.

While the CEQ did provide a definition and presented a basic framework for use and development of EAs, much less regulatory direction has been formulated by either CEQ or specific agencies faced with NEPA. Though the specific regulatory direction developed by CEQ and specific agencies has resulted in at least a small degree of standardization among the pool of EISs published each year, the much more enormous pool of EAs produced over the years has run the gamut from one-page technical memoranda to long documents that read (and weigh) much like an EIS.

The need for a book like this one is therefore critical if preparers of EAs are to receive the level of direction long enjoyed by preparers of EISs.

As many agencies have increasingly discovered the relative ease and reduced cost of preparing EAs rather than EISs, many have increasingly relied on EAs to cover actions that might have previously been addressed by an EIS. In some cases, the agencies were right in seeking the efficiency of an EA to enable sound environmental decision making without the burden of a full-blown EIS. In other cases, they mistook the purpose of an EA, which is to document a lack of potentially significant environmental impacts, not to masquerade as a perfunctory EIS; many viewed the EA as means of circumventing the transparency and public involvement objectives inherent to NEPA's objectives.

As agency budgets continue to become tighter in upcoming years, and the public continues to demand ever greater environmental protection and government transparency, the role of the EA in the NEPA process is only expected to become even greater than it is already. The need for a book such as this is sure to increase. *Preparing NEPA Environmental Assessments: A User's Guide to Best Professional Practices* serves to update and expand upon the previous guide to EAs, ensuring that the book continues to serve its evermore important function in the face of rapid change in government policy. The EA has in many respects come to dominate the day-to-day implementation of NEPA and, as we all seek more from government at less cost, may well come to be the dominant force of NEPA—eclipsing the EIS. This book will help NEPA practitioners transition to this future.

Introduction*

An act of historic proportions, the National Environmental Policy Act (NEPA) of 1969 pronounced the world's first environmental policy for protecting the state of the environment. At the time, few foresaw that NEPA would become a model for environmental policies adopted by nations the world over. With upwards of 500 environmental impact statements (EISs) and 50,000 NEPA environmental assessments (EAs) prepared annually, the influence of NEPA on federal planning is pervasive. Although EISs are generally prepared for larger and more controversial projects than are EAs, and EISs are therefore the NEPA document most familiar to the general public, the sheer number of EAs indicates that these shorter and less controversial documents also play a key role in the NEPA process.

Avoiding problems and fatal flaws

Before proceeding into a description of the National Environmental Policy Act environmental assessment process (EA), it is instructive to learn from and avoid past NEPA flaws and failures. NEPA was designed to force public officials to consider environmental effects before a final decision is made to pursue an action. Essentially, NEPA is intended to force decision makers to "look before they leap." At best, a problem-plagued NEPA process may lead to mistrust among the public, and lawsuits. For instance, in some years, the U.S. Forest Service lost a higher percentage of lawsuits than any other agency. At worst, a problem-plagued process can lead to misdirected or even dangerous decisions.

The recent tragedy of the Japanese Fukushima Daiichi power station highlights the importance of performing an objective, thorough, and rigorous assessment of proposed actions. In that case, mismanagement and faulty assessments had much to do with the partial meltdowns of three nuclear reactors that spewed radiation across a sizable portion of Japan.

---

* The views expressed by the authors are solely their own and do not represent those of any professional organization or employer, past or present.

The following example illustrates how a seriously misdirected planning and assessment process may have already led to flawed decisions with potentially devastating consequences. The principal lessons learned from this example are as applicable to the preparation of NEPA environmental assessments as they are to environmental impact statements (EISs). It is hoped that the reader can learn from this troubled process so that other equally dangerous decisions are not repeated.

A history of troubled nuclear licensing

NEPA became law in 1970. As an "independent agency," the U.S. Nuclear Regulatory Commission (NRC) has the distinction of being the only agency that continues to publicly claim that it is not subject to NEPA; in fact, the agency has claimed that it only complies with NEPA to promote goodwill and "because it chooses to do so." This is in keeping with its long record of deficient NEPA compliance. The Atomic Energy Commission (AEC), NRC's predecessor, had the distinction of triggering the first major NEPA lawsuit. This suit involved the Calvert Cliffs nuclear power plant. When sued in 1970, the former AEC claimed it did not have to comply with NEPA. In ruling against the agency, the judge admonished agency officials for their arrogance and disrespect for the law. The agency did manage to make one noteworthy contribution to NEPA compliance; the Calvert Cliffs lawsuit firmly established NEPA as "the law of the land." Unfortunately, as detailed below, substandard practices have carried over into the 21st century. The reader is referred to Chapter 4 for details of the Calvert Cliff's case.

In reality, the NRC and its predecessor, the Atomic Energy Commission (AEC) have a long history of proposing dubious and potentially dangerous projects. In the early 1960s, the AEC proposed to blast a harbor along the Alaskan coastline using nuclear bombs! Some members of Congress were aghast at the sloppiness of this proposal and the tremendous potential for devastating human and environmental consequences. On the coattails of this embarrassing project, Congress finally did something that lacked precedent. It ordered the agency to prepare an environmental report for the proposal. This project has since been criticized as potentially one of the most environmentally catastrophic ever proposed. Facing public uproar, the agency ultimately cancelled the project, in large measure because of the results of this study. Since then, the study has been called the world's first *de facto* EIS.[1] For more information on the landmark Calvert Cliff's lawsuit, and the AEC proposal to blast an Alaskan harbor with nuclear bombs, the reader is directed to the companion books, *NEPA and Environmental Planning* and *Global Environmental Policy*.[2]

By any measure, nuclear reactors are among the most perilous technologies in the world. The existing fleet of U.S. nuclear reactors was

Introduction xvii

originally licensed for a 40-year operating period. These licenses are now expiring or nearing expiration. These reactors are based on antiquated technological designs. The NRC's Division of License Renewal (DLR), directed by Brian Holian and Melanie Galloway (Deputy Director), is responsible for re-licensing (i.e., "license renewal" or LR) the entire fleet of aging nuclear reactors. The purpose of the LR process is to renew the current operating licenses for an additional 20-year period, thus extending the operating window to a full 60-year period; there is even talk of extending the operating period to 80 years or more.

What concerns critics and much of the public is that the NRC is performing a fast-paced and carefully choreographed process designed to show that the potential impacts of re-licensing aging nuclear reactor are essentially benign. And this charge is reinforced by the following fact. Never—not even in a single case—has one of these aging, antiquated reactors been denied a renewed operating license. As one NRC staff member begrudgingly acknowledged, "No licensing application has been rejected, and I will be surprised if one is ever rejected."[3]

How poor management and morale affect public safety

The NRC prides itself on being an "independent agency." For those unfamiliar with this term, an *independent regulatory agency* is a regulatory agency that considers itself independent from other branches or arms of the government. As described above, the NRC has gone so far as to argue that it is not even subject to NEPA, the law of the land, and only complies "…because it chooses to do so." Some of this problem can be traced to DLR and senior NRC management. Bo Pham has managed DLR's project branch (RPB1), which prepared many of the safety evaluations and EISs for re-licensing these aging reactors. DLR and RPB1 management have been plagued by morale problems and even its own project mangers (PMs) have complained of management-, environmental-, and safety-related issues. DLR staff members report that the DLR experienced the highest turnover rate of the entire NRC.[4] The DLR morale and management problems became so significant that a decision was eventually made to hold employee focus group meetings with the staff and project managers (PM) to determine the root cause of these problems. The DLR PMs were not shy about voicing critical and sometimes scathing comments during their focus group meeting.[5] But particularly disconcerting were comments such as[6]

- DLR "Managers don't listen—they act like know-it-alls."
- DLR "Managers are arrogant."
- DLR Managers are "bypassing the regulatory process and compromising the safety mission to impress upper management."

- "Poor management decisions" are being made.
- DLR is "sacrificing quality for schedule."

The focus group results were so scathing that NRC management hushed up the results. The final report was never circulated to the public or NRC staff. Perhaps more troubling, these unsettling comments were not lodged by outside antinuclear critics, but by the very PMs responsible for preparing the safety evaluations and EISs for reissuing the nuclear plant licenses. What follows only deepens these concerns.

The DLR managers play a vital link in the LR quality assurance process. For example, both Holian and Pham were responsible for reviewing these EISs and the safety evaluations, and for signing off on their accuracy, rigor, and thoroughness. But if DLR managers don't listen, act like know-it-alls, bypass the regulatory process and compromise the safety mission to impress upper management, make poor management decisions, and sacrifice quality for schedule, who is going to act as the critical stopgap to ensure that a license application has been properly investigated and fully vetted?

When schedules trump public safety

But it is the last of the DLR management failures ("sacrificing quality for schedule") that may be of most concern to critics. DLR management has mandated that all licensing applications are to be completed within an arbitrarily established schedule of 18 months. With only one technical exception, PMs are scrutinized on their ability to complete their re-licensing project within this capricious 18-month window, regardless of the complexity of the issues, environmental and safety concerns, or public sentiment. Senior-level NRC management routinely monitors the progress of these project schedules.

Now consider DLR's management practices in light of the Japanese Fukushima nuclear disaster. Japan's experience clearly illustrates what can happen when schedules and sloppiness trump quality, environmental concerns, and safety issues. As witnessed in subsequent chapters of this book, this is exactly what Congress was trying to avoid when it passed NEPA, i.e., prevent schedules, funding, and political factors from taking precedence over environmental and safety concerns. Congress's original goal has worked in many agencies, but obviously failed in others.

When NEPA becomes a charade

NRC prepares an EIS (technically a supplemental EIS) for each license renewal it grants for an aging reactor. The EIS considers alternatives to license renewal; however, the public is often bewildered to learn that none of these alternatives, not even the no-action alternative, have ever been given serious attention by NRC decision makers. Chapter 5 of the LR EIS investigates "severe accidents" of a nuclear accident; an episode that many experts consider to be one of the most catastrophic events that could result from any U.S. technology. The EISs produced under Pham's direction have routinely concluded that the risk from sabotage and beyond design-basis earthquakes (earthquakes the reactor is not designed to withstand) at existing nuclear power plants is "small." With respect to the assessment of a severe accident, Chapter 5 routinely concludes that, "The probability weighted consequences of atmospheric releases, fallout onto open bodies of water, releases to groundwater, and societal and economic impacts from severe accidents are small for all plants." This terse statement is all the stakeholders and public get. Perplexed? Not a word of consideration given to the crushing impacts that could result from a tsunami such as the one that crashed into the Japanese Fukushima reactors. Not a single utterance given to other potential incidents.

So how does DLR management justify the assignment of a "small" impact to a potentially catastrophic nuclear meltdown? It employs a mathematical "trick." RPB1 management cleverly "hides" potentially catastrophic consequences using something akin to mathematical smoke and mirrors. DLR takes the probability (which NRC argues is small), multiplies it by the consequences (assume it is large), and then concludes that the human, environmental, and socioeconomic impacts are small because the probability is so small. But let's step back for a moment and reframe the context of this problem. The EIS runs for hundreds of pages, examining every conceivable impact, from biota to air emissions and water usage. Then, when it comes to the real issue that everyone is concerned with—the true issue that lies at the heart of the entire licensing process—it provides nothing but a cursory dismissal of the potential impact and a scant conclusion that the consequences of a large-scale accident or meltdown would be "small." Obviously DLR and RPB1 management is going to great lengths (hundreds of pages) using the mathematical equivalent of "pulling a rabbit out of the hat" to hide the real issue that could kill the issuance of a renewed nuclear operating license.

Let's reconsider the context of the DLR's decision making process from yet another perspective. Are the impacts of a potential accident that could result in radiation deaths, birth defects, evacuation of thousands of downwinders, property damage in the hundreds of billions of dollars, and near-permanent contamination of hundreds if not thousands of square miles "small"? Were the potential consequences of Chernobyl, Three Mile Island, and more recently the Japanese Fukushima Daiichi power station disaster "small"?

If the probability of an accident is as small as RPB1 management claims, why have there been four other near-catastrophic accidents (near misses) in the United States, in addition to Three Mile Island? Clearly, since the U.S. has already experienced a partial meltdown of the Three Mile Island nuclear reactor and four other near misses, the probability cannot be that "small." So even DLR's mathematical trick of smoke and mirrors is indefensible. Again it is obvious that NRC management is going to great lengths, obscuring the facts, to avoid having to announce to the public and stakeholders that the impact of a nuclear accident could result in catastrophic human, environmental, and socioeconomic repercussions, as great as or perhaps even greater than that experienced by Chernobyl or Japan. Such egregious practice is a lawsuit waiting for a litigant.

Cumulative risk to the public hasn't even been considered

This brings us to the issue of cumulative impacts. In addition to gauging direct and indirect impacts, a NEPA analysis must also rigorously investigate the cumulative impact of an action (i.e., the combined impacts of other past, present, and reasonably foreseeable actions).[7] For decades, the courts have made it clear that cumulative impacts must be publicly disclosed and rigorously examined. DLR's EISs do address cumulative impacts. But what they conveniently ignore is the issue at the heart of the entire licensing process—risk. As we have just seen, these EISs do present a terse summary of the risk from potential accidents. But this analysis only considers the probability and risk of an accident from a single operating plant. But this is not the actual case as there are over 100 operating reactors in the U.S. The cumulative risk to the American public from an entire fleet of operating reactors has been utterly disregarded. While Chapter 5 of the generic EIS on which specific plant licenses are based mentions the word "cumulative" twice, no attempt was made to determine cumulative risk or cumulative impacts of an accident in terms of a standard cumulative effects assessment. These EISs are therefore defective, as they have not computed the total or cumulative risk and effect of an accident from over 100 operating reactors.

That the cumulative impacts from a major accident would be felt by millions and could sweep across many states if not much of entire continent is undeniable; radiation released from a single accident could spread across the U.S. potentially endangering millions of Americans; then there are the socioeconomic impacts and losses that would be felt across the entire continent. The actual cumulative risk that members of the public are exposed to is much higher than that described in DLR's EISs. The RPB1 manager dismissed this analytical defect when it was brought to his attention. It should come as no surprise that this is a lawsuit waiting to happen.

Futile paper exercise

The NEPA implementing regulations state that the section on alternatives is the "heart" of the NEPA analysis.[8] There is one, and only one, fundamental purpose for preparing an EIS: to evaluate *alternatives and mitigation measures* that *can be implemented* to avoid or reduce potential impacts. RPB1's assessments provide only a cursory review of alternatives and mitigation measures—just enough so that the decision maker can claim they weren't overlooked. The assessment is probably insufficient to withstand a serious legal challenge. But from this point, the defensibility of the alternatives assessment only deteriorates.

While alternatives and mitigation measures are given superficial treatment, the simple truth is that DLR has *never* seriously entertained an alternative or declined a license renewal application (i.e., adopted the no-action alternative, which would involve denying a renewed operating license) in favor of one of the alternatives. Likewise, it does not compel, oblige, or even attempt to convince applicants to adopt mitigation measures identified as part of the EIS re-licensing process. NRC essentially ignores alternative courses of action despite the fact that courts have long held that reasonable alternatives must be seriously considered during the final decision making process.

In terms of considering alternatives, NRC takes the stance that it lacks authority to compel an applicant to implement a mitigation measure or alternative identified in the EIS; although this excuse is debatable, it is also clear that the NRC does not even attempt to seek authorization from Congress to compel an applicant to consider a superior alternative or mitigation measure. NRC has taken this stance even though it is clear that one of the goals of NEPA is to prepare an objective analysis that informs policymakers so that they can change statutes allowing an agency to pursue alternatives and mitigations measures for which they lack authority.[9]

When the no-action alternative isn't even afforded serious consideration

Perhaps of most concern is that the no-action alternative has never been seriously entertained. Approvals are virtually rubber stamped on a near assembly-line basis. Even DLR staff has openly and publicly admitted that NRC does not seriously consider alternatives beyond the option of re-licensing a nuclear power plant. Consider what one of Holian and Pham's own project managers in charge of re-licensing a nuclear reactor had to say at a NEPA public meeting on a draft EIS to renew an operating license. When asked about the choice of taking no action and shutting down the reactor on or before its operating license expired, this project manager publicly stated, "...that option *wasn't even considered* because of the important role which Cooper Nuclear Station plays in providing energy."[10] Not even considered! In other words, the project manager in charge of re-licensing the reactor openly admitted in a public meeting that the no-action alterative (in addition to all the other alternatives) would not even be considered by the final decision maker? The NEPA implementing regulations state that a final decision is not to be made until the final EIS has been issued. Yet, the project manager publicly admitted that the decision to renew the operating license had essentially been made because of the perceived need that the "...Cooper Nuclear Station plays in providing energy." For some forty years, the courts have made it crystal clear that agencies must seriously consider the "...adoption of all reasonable alternatives." They have also made it repeatedly clear that the no-action alternative may not be ruled out and that a final decision may not be made until the NEPA process has been completed. Apparently, the DLR has yet to get the message. Again this is a misstep ripe for a legal challenge.

Neglecting to consider realistic alternatives beyond renewing the operating license

Added to this is the fact that DLR intentionally skews the alternatives in such a way as to make them appear unreasonable. For instance, many of its EISs provide nothing but a scant one-paragraph description of the alternative of constructing a new reactor based on a modern design that would be much safer compared to the antiquated designs that are being relicensed. In most cases, the New Reactor alternative is dismissed because it could not be completed in time to meet the expiration date of the operating license. This reasoning is bogus. For instance, DLR totally neglected to consider the very realistic option that the operating license might be extended for a period sufficient to allow the present nuclear operator to construct a new and safer reactor; this would allow the present reactor to continue operating for a defined period and would force the operator to either replace the existing reactor with a new and safer one, or shut down. So it goes for most of the alternatives that are "examined."

So if the alternatives, including the no-action alternative are not even considered, what is the point of going through all the time, money, and charade in preparing the NEPA assessment? The answer is simple: To appease the public and give the appearance of complying with NEPA's public participation mandate, thereby reducing the risk of a lawsuit. Paradoxically, such ill-conceived practices have set up a near perfect basis for a lawsuit.

In fact, DLR management has routinely reached its decision to relicense a reactor based on gawky and awkward logic such as the "...impacts of license renewal are not so great as to make license renewal unfavourable" as justification for reaching the decision to relicense an aging reactor; apparently, a catastrophic nuclear power plant accident such as that recently experienced by Japan is "not so great" as to cause the DLR to reconsider the wisdom of relicensing some aging reactors.

This brings us to yet another point. One of the principal purposes of NEPA is to inform the public and allow them to provide input into the assessment process. For each of its relicensing efforts, DLR does hold a public scoping meeting and a meeting to receive comments on the draft EIS. However, a careful review of public comments shows that most of this input is rejected, often based on the fact that NRC argues that the comment is outside the plant's "licensing basis." Critics can find no basis for an agency rejecting a comment simply because it lies outside a plant's licensing basis. For example, one member of the public raised a concern that a solar flare could destroy the plant's cooling capability, resulting in a meltdown. The response was to respectfully dismiss the consequence without any serious evidence that such an event could be properly mitigated. Using such logic the DLR could dismiss the potential effects of a Japanese-like Fukushima tsunami based on the fact that such an event does not fall within the plant's licensing basis. And in fact, they routinely dismiss comments such as this. This is particularly disconcerting given the fact that Japanese engineers had dismissed concerns about the potential of a gigantic tsunami striking its Fukushima Daiichi nuclear reactors, with a simple memo.

Such condescending practices defeat the entire purpose of NEPA—to protect the environment and public health. One is left to ponder what value, if any, do such public comments contribute in terms of safeguarding the environment or public safety? When everything is said and done, DLR will have relicensed an entire national fleet of aging and antiquated reactors, never having seriously entertained any other option. Yet in doing so, DLR has laid the foundation for successful legal challenges.

To be fair, such poor practices are not a reflection on DLR's project managers or staff specialists. As we have already seen, these flaws can be traced to poor management practices. Excellence comes by example and from the "top down." A staff cannot be expected to produce quality decision making assessments if the management is not truly committed to NEPA's intent and goals. If a ship sailing in the wrong direction, perhaps a new captain is needed to turn it around.

A re-review of license renewal

As just witnessed, DLR's NEPA process is the equivalent of a futile "paper chase," carefully choreographed to hoodwink the public and Congress into believing that most of the impacts, including those of a severe accident are "small." Rather than use NEPA as a true scientific assessment process for publicly and forthrightly evaluating impacts and comparing alternatives, NRC simply approaches it as another hurdle to jump in its accelerated mission to relicense the nation's fleet of antiquated reactors.

Given such mismanagement practices, what does this behavior say about nuclear plant licenses already renewed? Do stakeholders fully appreciate the extent to which most public concerns and comments are casually dismissed with terse rationalizations? Do they realize that DLR management merely meanders through a pointless exercise of evaluating alternatives and mitigation measures, and that these options are not even afforded serious attention or consideration? Does the public understand that the true cumulative risk was never even examined in detail or publicly vetted, and that they actually face much greater risk of an accident than each individual EIS would lead them to believe? How will public opinion be affected when people come to realize that the true risk of a severe nuclear accident is actually "large" or "catastrophic" rather than "small"? Would such knowledge have affected their support or opposition to the renewal of nuclear plant licenses in the first place?

The European Union (EU) has taken the Japanese lesson to heart. The EU has already ordered its member states to perform a rigorous and comprehensive reassessment of all nuclear power plants. These assessments must consider each plant's vulnerability and ability to withstand a full spectrum of potential risks and hazards—from earthquakes and floods to plane crashes and even terrorist attacks. Yet, NRC continues to drag its feet, arguing that it is an "independent agency" and steadfastly refusing to seriously address public NEPA comments for reasons such as "the comment is not within the plant's 'original licensing basis.'" Likewise, NRC has refused to evaluate various scenarios in its NEPA analyses, such as terrorist attacks, unless a federal court within the affected jurisdiction has already ruled that it must do so. It continues to deny that the risk of a severe accident is "large," instead maintaining that it is "small." Consistent with the EU, the time has come for NRC to open up its entire review process and perform a full reexamination of all license renewal EISs and safety assessments issue to date.

Dealing with significant new circumstances or information

For years, NRC's "mathematical magic" duped some of the media and public into believing that the risk of a nuclear accident was "small." In light of the Japanese experience it is now clearly evident that the probability and impacts of a severe accident are not only real but can be devastating; based on the Japanese disaster and the significant new circumstances and information that have now been revealed, no reasonable person, including a federal judge, can conclude otherwise. Both the NRC's and the Council on Environmental Quality's NEPA implementing regulations

(40 Code of Federal Regulations [CFR] 1502.9) provide clear, specific, and unambiguous direction regarding what must be done when an agency finds:

> There are significant new circumstances or information relevant to environmental concerns and bearing on the proposed action or its impacts.

In all such cases, the agency "shall prepare" a supplemental EIS to evaluate the proposal in light of the *significant new circumstances or information*. There are no exceptions. The Japanese experience clearly demonstrates that both *significant new circumstances* and *information* now exist.

Reexamining and supplementing every license renewal issued to date

This significant new information not only affects license renewal EISs now being prepared, but perhaps more importantly, those that have already been approved. As required under 40 CFR 1502.9, a supplemental EIS must be prepared for each and every license renewal granted to date. Not only are supplemental EISs required, but the NRC will finally have to publicly assign a finding of "large" or perhaps even "catastrophic" (in lieu of "small") to the risk of a severe accident. This also will require performing a comprehensive reassessment of alternatives and mitigations measures to accompany the new finding of a "large" impact to severe accidents. Tsunamis and flooding are just some of the potential hazards that will have to be rigorously reassessed. This will be no small feat. In the meantime, the status of operating plants that have already received renewed licenses will remain in limbo.

Sooner or later the public and watchdog organizations will begin to realize how safety and accident issues have been twisted, obscured, and compromised. In the end, it may require a lawsuit to force NRC into implementing NEPA as it was originally intended—a true planning and decision making process rather than a fast-track "document preparation" charade. Legal action may be required to force it to reexamine and supplement every license renewal issued to date. Alas, it may even require a suit to force NRC into seriously, frankly, and publicly considering harsh mitigation measures or even the possibility of refusing a renewed operating license to some plants that pose particularly grave threats. It will then be a matter of time before Congress begins asking some very difficult and embarrassing questions.

In trying to shed the best light on this sad state of NEPA affairs, NRC will, of course, deny that the license renewals issued to date are defective.

In light of the Japanese fiasco, NRC is already flaunting its efforts to reexamine some hazards such as earthquakes. But this does nothing for the majority of plants that have already received renewed operating licenses based on faulty and flawed NEPA analyses; and why did it take a Japanese fiasco to prompt NRC into finally reassessing such issues when the public had submitted NEPA comments for years demanding that such issues be examined? All of this could have been avoided if Holian and Pham had simply been diligent and candid about the true impacts as they were preparing license renewal EISs and accompanying safety assessments.

The future of nuclear risks

An "assembly line" process designed to rubberstamp operating licenses for antiquated reactors as if they were widgets, on the shortest possible schedule, will ultimately leave major questions for a society that must struggle and live with the long-term consequences. Jeffrey Loman, deputy regional director for the Bureau of Ocean Energy Management, Regulation and Enforcement (formerly the Minerals Management Service [MMS]) stated that prior to the Deepwater Horizon disaster, the MMS had come to a belief that it had a "gold-plated" safety system—a belief that led to dangerous levels of complacency. There are striking parallels between the mindset of the NRC and former MMS management; like NRC, the MMS had also concluded that the probability of a severe accident involving the Deepwater Horizon project was "small." The tragic lesson of the Deepwater Horizon oil spill disaster is simple—accidents and calamities having grave consequences *can* and *do* occur. Has NRC management likewise relicensed a ticking time bomb—the equivalent of a Japanese Fukushima Daiichi nuclear reactor meltdown?

To date, NEPA's contribution to DLR's decision-making process is nil. The taxpayer and the public would be better served if DLR simply petitioned Congress for a waiver from the requirement to prepare EISs for relicensing the aging fleet of nuclear reactors. Meanwhile, a catastrophic meltdown may be a week away or 10 years into the future. Some experts warn that the impacts from a major accident could be so severe that a major city or a good chunk of an entire state could be permanently "lost." Every day an aging reactor ticks away with a renewed license, granted as part of a flawed NEPA process puts the public that much closer to a U.S. version of Fukushima.

NEPA provides a rigorous and systematic mechanism for safeguarding the public and our environment—but only if public officials are committed to honest, objective, and rigorous assessments that truly inform policymakers and the public of the costs and risks. As Lynton Caldwell, the Father of NEPA often declared, the goal of safeguarding our environment and public health can only be assured when an organization is

Introduction

truly committed to such goals. No safety or assessment process, not even NEPA, will make up for a façade, or an ill-conceived, or slipshod nuclear licensing process, nor can it completely compensate for poor management decisions. However, when combined with an open, objective, and serious planning and assessment process, NEPA provides a vital tool for mitigating potential hazards that may befall ill-conceived projects. It is to this end that this book has been written and dedicated.

Goals of this text

This text has been specifically designed to bridge this regulatory chasm. Applying the "Rule of Reason," this text identifies and describes relevant EIS regulatory requirements that can be logically interpreted to also apply to preparation of EAs. This text also draws on the professional experiences and assimilates best professional practices from seasoned practitioners who have spent years preparing EAs. The goal is to provide the reader with a reasonable, definitive, consistent, and comprehensive methodology for managing, analyzing, and writing EAs.

The limited direction that has been propagated by the CEQ for preparing EAs is scattered throughout NEPA's implementing regulations, executive orders, guidance, and case law. To date, no text has collectively compiled and synthesized this information into a single source. This book is designed to provide the reader with a single, integrated, and comprehensive source of the relevant guidance and requirements. Diverse sources have been drawn from, including the CEQ NEPA regulations and guidance documents, executive orders, professional papers, and experience of NEPA practitioners. Case law has also been integrated throughout this book to provide additional direction or clarification. Finally, the National Association of Environmental Professional's NEPA Working Group has provided a valuable source of information, experience, and expertise from which to draw.

Objectives of this text

This text is unique in that it:

- Provides the user with the most comprehensive and thorough description of the EA process written to date. Ultimately, the goal is to provide the reader with a reasonable, definitive, consistent, and comprehensive methodology for managing, analyzing, and writing EAs.
- Comprehensively describes the step-by-step process for managing and preparing an EA. An approach for evaluating environmental impacts is also described.

- Details all documentation requirements that the EA must meet. Additionally, recommendations are offered for promoting a more defensible assessment as well as improving the goal of excellent environmental planning. Lessons from case law are integrated with the relevant requirements.
- Addresses the regulatory vacuum under which practitioners have struggled with the paradoxical problem of preparing publicly defensible EAs given only minimal regulatory direction. The text identifies and describes relevant EIS regulatory requirements that can logically be interpreted to apply to preparation of EAs. The experiences of seasoned practitioners and best professional practices are assimilated to provide the reader with a comprehensive, definitive, and defensible methodology for preparing EAs.
- Addresses problems and dilemmas that have traditionally plagued preparation of EAs. Specific tools and approaches are suggested for resolving such problems. Emphasis is placed on introducing methods and procedures for streamlining the EA process.

Audience

This text is designed for beginners and experts alike. It begins with the fundamentals and advances into increasingly more advanced subject matter. Experienced practitioners can use the book as a resource for quickly reviewing issues, or as a comprehensive textbook. Although primarily aimed at professionals in government, consulting, and the private sector who prepare and review EAs, this book also lends itself to individuals who seek only an introduction to certain selected topics. Individuals and groups include decision makers, analysts, scientists, planners, regulators, project engineers, and lawyers, to name just a few. People in advocacy or citizen groups who seek to challenge an NEPA compliance action will find the book equally useful. The book can be used by university students in environmental curricula and by instructors who teach professional courses.

While this book provides the reader with guidance for preparing environmental assessments, the author stresses the importance of consulting with environmental scientists and engineers, NEPA specialists, regulatory analysts, and legal experts, particularly in areas involving complex or controversial issues. The reader is advised to consult the actual regulatory provision for the details and precise wording.

References

1. O'Neil, D., "Project Chariot: How Alaska Escaped Nuclear Excavation," *The Bulletin of the Atomic Scientist* (December 1989): 35.
2. Eccleston, C., *NEPA and Environmental Planning: Tools, Techniques, and Approaches for Practitioners*. CRC Press 2008, 450 pages.
3. Privileged Communication with NRC staff. Names withheld (2010).
4. Ibid.
5. Internal DLR project managers focus group meeting, held September 14, 2010.
6. NRC internal report regarding results of DLR focus group meeting that was held on September 14, 2010. Also includes supplemental statements supplied by DLR project managers that attended the focus group meetings.
7. 40 Code of Regulations 1508.7.
8. 40 Code of Regulations 1502.14.
9. CEQ, Forty Most Asked Questions, 2b.
10. *The Nemaha County Herald*, "Only Positive Remarks Presented Regarding Cooper Nuclear Station's License Renewal" 15 April 2010, http://www.anews-paper.net/index.php?option=com_content&view=article&id=354:only-positive-remarks-presented-regarding-cooper-nuclear-stations-license-renewal&catid=1:local&Itemid=2
11. Eccleston, C. and March, F., *Global Environmental Policy: Principles, Concepts and Practice*, CRC Press (Lewis Press), 412 pages.

section one

NEPA *environmental assessment process*

chapter one

Overview and historical development of NEPA

> Do the right thing. It will gratify some people and astonish the rest.
>
> —Mark Twain

1.1 Introduction

Prior to the National Environmental Policy Act of 1969 (NEPA), an agency of the federal government could make a decision, hire a contractor, and bulldozers could show up on public land two weeks later. Since its passage, NEPA has forced the federal government to "look before it leaps." The environmental assessment (EA), the subject of this book, is one of several types of documents prepared by agencies of the United States government under NEPA. Other common NEPA documents are the environmental impact statement (EIS), categorical exclusion (CATEX), and various comment response and decision documents. NEPA is an outgrowth of a dramatically increased national interest in environmental planning and conservation that intensified in the years following World War II, especially during the 1960s. As described in the following section, the environmental movement in the United States—manifested today through NEPA and a suite of other federal environmental regulations such as the Clean Air Act, Clean Water Act, and Endangered Species Act—began to take root well over a century before enactment of NEPA. An understanding of the EA and other elements of modern NEPA practice requires an understanding of the roots of American environmental planning and practice.

Section 1.2 discusses the history of environmental protection and what we currently recognize as the "environmental movement." Sections 1.3, 1.4, and 1.5 provide a brief overview of NEPA, its history, statutory language, and regulations, with a special focus on the EA.

1.2 Origin of environmental movement

The modern environmental movement, has primarily been a post-industrial phenomenon initiated mainly by the United States and

Europe. While the environmental movement in the United States took off in the 1960s, its origins can actually be traced back to the second half of the nineteenth century as an effort to save the nation's wildlife heritage. Interest in the preservation of natural lands has even deeper roots in Europe, although conservation interests up through the nineteenth century largely focused on the preservation of hunting opportunities for the nobility. Many of the preserved natural spaces in European cities, such as Hyde Park in London or the Bois de Boulogne in Paris, resulted from conversion of former royal hunting preserves into environmentally pleasing recreational spaces for the growing urban middle class resulting from the Industrial Revolution. The same thinking led to the establishment of Central Park in New York City, similar natural spaces in other American cities, and the magnificent system of national, state, and regional parks, forests, preserves, and other preserved natural spaces that permeate the American landscape today.

1.2.1 Nineteenth century

The origins of the modern environmental movement in the United States can be traced in part back to concerns for American bison, which had been hunted to near extinction. Observation of and concern about spectacular declines of other wild fauna constitute the roots of what would eventually lead to the Endangered Species Act, a federal act passed shortly after, and often viewed in association with NEPA. The 1864 publication of *Man and Nature* by George Perkins Marsh also contributed to this fledgling movement. Marsh wrote about environmental degradation and promoted natural healing of damaged environments.

One of the earliest American environmental successes involved the 1872 enactment of the law establishing Yellowstone National Park, the world's first national park, thereby setting a precedent for the preservation of scenic federal lands. The very concept of a "park" ultimately reflects the "parks" or "enclosures" of naturally vegetated lands in Europe established for hunting by nobility. In 1873, the American Association for the Advancement of Science petitioned Congress to halt the unwise use of natural resources. In the following years, Congress continued to lay the foundation for federal protection of lands by expanding the national park system, establishing national forests and the U.S. Soil Survey. Many states followed suit with the establishment of state parks and state forests. In 1891, John Muir founded the Sierra Club, one of the seminal organizations of the environmental movement, which has remained active ever since.

1.2.2 Early twentieth century

As a result of his own love of the outdoor environment, and influenced by the work of Marsh and Giffort Pinchot (the first chief of the U.S. Forest Service), President Theodore Roosevelt set aside 125 million acres of federal lands for protection during his term of office (1901–1909). To prevent vandalism at prehistoric Indian sites in the Southwest, Congress passed the Antiquities Act in 1906, authorizing the president to establish national monuments on federal lands. Early federal actions on behalf of the environment, however, were not free of controversy. John Muir claimed that Roosevelt's policies served to stimulate economic uses of lands that should be protected in their original state. This criticism stimulated a debate over environmental management that continues to this day.

As important as they were, these early policies and statutes proved to be inadequate in the face of events to come. The economic depression beginning in 1929 initially overshadowed the growing interest of the increasingly affluent American public in preserving green spaces for recreation in the 1920s, but conservation subsequently became a key benefactor of President Franklin D. Roosevelt's New Deal. Roosevelt's Civilian Conservation Corps put unemployed young men to work planting trees on eroding soils and constructing recreational facilities in national parks. During the 1930s large parts of the Southwest experienced drought and a series of environmental calamities that created economic havoc and induced large-scale migration from Oklahoma and other locations westward. In response to these events, the Roosevelt administration created the Soil Conservation Service and the Agricultural Stabilization and Conservation Administration to promote a version of what we might today call "sustainable agriculture" through soil conservation and other beneficial land management practices.

Though the early roots of the environmental movement were established, national concerns during the 1940s and 1950s focused on winning World War II and the reconstruction of shattered European economies in the postwar era. The environmental movement had yet to move beyond land management and species conservation to deal with the larger issues of pollution and the environmental impacts of an industrial society.

1.2.3 Environmental decade of the 1960s

Momentum for a comprehensive environmental policy fermented through the 1960s. To fully comprehend the forces that led to the enactment of a national environmental policy, one must appreciate the context in which NEPA was created. The American public and Congress alike were becoming increasingly concerned that the environment was deteriorating at an alarming rate. This was an era characterized by the publication of Rachel

Carson's *The Silent Spring*, the Santa Barbara oil spill, and the Love Canal incident. Lake Erie was pronounced "dead," and smog alerts were issued in major cities across the nation. The Bureau of Reclamation was proposing to build a dam on the Colorado River that would flood the Grand Canyon. There were visibly polluted waterways, blighted urban landscapes, and unprecedented expansion of sprawling suburbs over the pastoral landscapes most visible to urban Americans. Events of this scope combined to stimulate an environmental activist movement. Perhaps no event captured the public's imagination more than the nightly news broadcasting scenes of the Cuyahoga River in Cleveland, Ohio, which was so polluted that it actually caught fire!

Leaders of the fledgling environmental movement aggressively promoted public awareness throughout the 1960s and created a political lobby to promote enactment of new laws, including NEPA. While the first version of the Clean Air Act was enacted in 1955, the larger body of environmental law encompassing protection of land and water media, as well as far more comprehensive protection of the air, remained an ideal for the future, to be realized only after the enactment of NEPA.

1.3 Historical development of NEPA

Many avenues were available to Congress for addressing the nation's looming environmental problems. Congress could have dealt with the impending environmental quagmire by amending the laws authorizing federal programs, one statute at a time, but chose not to. Instead, Congress adopted NEPA—a single statute for overseeing environmental degradation, at the federal level. In doing so, Congress gave priority to addressing environmental issues in the planning phase before they evolved into larger problems.

1.3.1 Birth of NEPA

As far back as 1959, Senator James Murray had proposed establishment of a council overseeing environmental quality. Support for a national environmental policy evolved slowly over a period of more than 10 years. Interestingly, prior to NEPA's enactment, there was actually a precedent for preparing a study of possible environmental impacts from proposed projects. In the early 1960s, the Atomic Energy Commission (AEC) was required by Congress to prepare an environmental report for a proposal to use nuclear explosives to blast a harbor along the Alaskan coastline. This project has since been criticized as potentially one of the most environmentally catastrophic ever proposed. The project ultimately did not go forward, in large measure because of the results of this study, which has been viewed as the first *de facto* EIS.[1]

Chapter one: Overview and historical development of NEPA

This trailblazing path would provide a model for NEPA in the later 1960s. Congress was increasingly hearing testimony from the scientific community regarding the alarming rate of environmental degradation and the potential for disaster. In response to growing environmental consciousness, the concept of NEPA was originally expressed in a proposed policy statement crafted by Lynton Caldwell, special assistant to the Senate.[2] For this reason, Caldwell has been called the father of NEPA.

The Senate Committee on Interior and Insular Affairs reported:

> ... that in spite of the growing public recognition of the urgency of many environmental problems and the need to reorder national goals and priorities to deal with these problems, there is still no comprehensive national policy on environmental management. There are limited policies directed to some areas where specific problems are recognized to exist, but we do not have a considered statement of overall national goals and purposes.[3]

Senator Henry Jackson of Washington State was particularly concerned about issues such as timber-cutting in his home state's forests and spills from oil tankers entering Puget Sound. Following Jackson's contentious committee hearing on a Bureau of Reclamation proposal to dam the Colorado River above the Grand Canyon, he recognized that a mechanism was needed to force federal agencies to instill environmental considerations into their decision making process. His leadership in the Senate, together with that of Representative John Dingell of Michigan in the House, eventually led to enactment of NEPA. Individual drafts of the act were prepared by both the House of Representatives and the Senate. The Senate's draft version of the bill was the more comprehensive, but lacked a clear vision of a national environmental policy.

NEPA was modeled, to a great extent, after another trailblazing statute, the National Employment Act of 1946, which established a Council of Economic Advisers to assist and advise the president on economic matters.[4] Rather than create another new and expansive bureaucracy or re-engineer the existing federal agency apparatus, Congress wisely chose to craft the national policy by supplementing the existing statutory charter of federal agencies. From that point on, agencies would be expected to balance the goal of preserving the environment with other competing factors and policies, such as economic growth. In short, agencies would be required to infuse NEPA into their traditional decision making processes. Congress created a Council on Environmental Quality (CEQ) to administer the act, and each agency was to assume responsibility for self-enforcement, guided by the CEQ.

Some senators and congressmen argued that an environmental policy statute alone would not be enforceable and proposed drafting an environmental amendment to the U.S. Constitution. Other supporters, recognizing that powerful business interests would oppose environmental restrictions on the private sector, urged a statute that would overcome industrial opposition by focusing exclusively on the actions of the federal government. They believed—correctly, as it turned out—that passage of such a bill would demonstrate the seriousness with which Congress viewed environmental protection and set a precedent for subsequent legislation that would control the environmental behavior of the private sector. The federal government, because it is the single largest entity in the United States and because of the vast scope and nature of its actions, accounted for a disproportionately larger share of the nation's environmental degradation. If for no other reason, such an act (even if limited to federal actions) would clearly have profound implications on future environmental disruption.

1.3.2 "Detailed statement"

The NEPA statute was the subject of considerable debate. Much of this debate centered on the bill's action-forcing mechanisms, which appeared in the Senate but not in the House version. There were also those who opposed even the limited action-forcing provisions. Senator Jackson was adamant that such a provision be incorporated to ensure that the act not be merely a paper tiger. In respect to an action-forcing mechanism, a compromise was eventually struck in a conference committee that included the following key provision:

> All agencies of the Federal Government shall ... include in every recommendation or report on proposals for legislation and other major Federal actions significantly affecting the quality of the human environment, a detailed statement by the responsible official (Sec.102(2)(C) of NEPA).

As debate continued, the draft act slowly evolved. One early version used the term "finding" instead of "detailed statement." Senator Edmund Muskie of Maine, successfully negotiated the substitution of "detailed statement," fearing the statute would be too weakly worded, making it easy for federal agencies to circumvent the act's intent. The detailed statement would have to be technically based, not merely an agency's statement of opinion or direction. This action-forcing mechanism would later become known as the EIS. Thus, Senator Muskie has been referred to as the father of the EIS. Such detailed statements were to include, among

Chapter one: Overview and historical development of NEPA 9

other requirements, a description of "alternatives to the proposed action" (Sec. 102[2][C][iii]). To foster objectivity, the conference committee also injected items promoting public involvement.

One school of thought argued that agencies should not only be required to consider alternatives to a proposed action, but should actually be required to select an environmentally benign alternative. However, such a provision was considered unduly restrictive and not practicable, and hence was resisted. Instead, it was successfully argued that NEPA's effectiveness would result from heightened awareness, which would ultimately lead to better environmental decisions. Consequently, in passing NEPA, Congress imposed no substantive requirement on agencies to select an environmentally benign or least damaging alternative (sometimes termed today the "environmentally preferable alternative"). Thus the clear intent of the act is to force agencies to openly consider the environmental effects of alternatives so as to heighten awareness and promote informed decision making, but not to actually tie the hands of agencies in implementing their missions.

1.3.3 Passage of NEPA

Agreement was reached on the final language of the bill. The NEPA statute received the unanimous vote of the Senate Interior Committee and enjoyed widespread support among members of Congress. The significance of this act was reinforced when President Nixon chose to sign NEPA into law on New Year's Day of 1970, proclaiming this as "my first act of the decade." Thus NEPA has the unique distinction of being the first law the U.S. enacted during the new decade of the 1970s. Upon its enactment, few congressional members foresaw the broad ramifications that NEPA would later have for federal decision making or that it would be a model copied by nations around the world. Following in the path of NEPA, Congress established the Environmental Protection Agency (EPA) in 1970. The world's first "Earth Day" was celebrated on April 22, 1970.

Perhaps NEPA's single greatest contribution has been that it requires federal agencies to consider environmental issues in reaching decisions, just as these agencies consider other factors that fall within their domain. The bill's champion, Senator Henry "Scoop" Jackson of Washington State, chief sponsor of NEPA in the Senate, declared that "no agency will be able to maintain that it has no mandate or no requirement to consider environmental consequences of its actions." In underscoring the historical significance of this act, Jackson stated that its passage would be:

> ... the most important and far-reaching environmental and conservation measure ever enacted ... more than a statement of what we believe as a people and as a nation. ... It serves a constitutional function in that people may refer to it for guidance in making decisions where environmental values are found to be in conflict with other values.[5]

1.3.4 NEPA's implementation

No environmental statute can effectively work in a vacuum that lacks either implementing guidelines or regulations. President Nixon granted the CEQ authority to issue NEPA guidelines in 1970.[6] The CEQ published interim guidelines the following month and final guidelines in 1971. These were revised in 1973 to provide more detailed instructions.

Following NEPA's enactment, agencies often took actions with total and open disregard for NEPA's requirements. Such disregard resulted partially from the fact that, unlike most other environmental acts, NEPA had (and still has) no criminal or civil penalties associated with noncompliance. The U.S. federal court system became the default mechanism for ensuring that the act was not merely the paper tiger that Senator Jackson sought to avoid. Citizens and organizations with legal standing may sue any federal agency for failing to enforce the provisions of NEPA in connection with a proposed action. Thus the act provides the environmental movement, including old-line organizations like the Sierra Club and the National Audubon Society and newer advocacy groups such as the Environmental Defense Fund and the Natural Resources Defense Council, with a mechanism for challenging environmentally questionable projects. In addition to these "environmental watchdogs," plaintiffs have included many citizens groups and even state and local government entities.

The years prior to issuance of CEQ's NEPA regulations were often a difficult time for agencies seeking to comply with NEPA. This was true even though CEQ, beginning in the early 1970s, had introduced a series of interim guidelines for preparing EISs, pursuant to an executive order signed by President Nixon. CEQ's intent was that these guidelines were to be considered nondiscretionary and binding. However, they were widely viewed by agencies as advisory. Moreover, they were seen as providing insufficient direction on how to comply without creating unnecessary delays and paperwork.

CEQ had offered little guidance on precisely what constituted a "detailed statement," although the guidelines did distinguish between a "draft" and a "final" statement. As a consequence, agencies were often challenged in court on the adequacy and level of detail contained in such statements. Some agencies, such as the Atomic Energy Commission (AEC), the forerunner to the Nuclear Regulatory Commission (NRC), openly showed contempt for NEPA's requirements. Early in its inception, beginning with the landmark case of *Calvert Cliffs*, in which the AEC was sued for refusing to comply with NEPA. The courts concluded that NEPA established "a high standard for ... agencies." Thus in the case of *Calvert Cliffs*, a precedent was firmly established that all agencies must comply with NEPA's action-forcing provision to prepare a detailed statement, unless it can be shown that this requirement directly conflicts with another statutory provision, a circumstance in which there are surprisingly few conflicts.

While early EPA court decisions strengthened NEPA's compliance requirement, they often ruled in favor of additional detail. Many EISs were already considered bloated, nonanalytical documents, sometimes exceeding one thousand pages in length; the courts only exacerbated the situation. As time passed, the NEPA process began to resemble a sailboat with a broken rudder. An early concern of the CEQ was that agencies appeared to view the requirement for an EIS as a perfunctory end unto itself rather than as a vehicle to foster sound, environmentally aware decision making. This concern persists to this day, despite increased efforts by CEQ, EPA, and various environmental activist groups (as well as many pro-growth and smaller government activist groups) to simplify the process of NEPA compliance without sacrificing its value as a planning process.

1.3.5 NEPA's implementation regulations

Complaints about NEPA were increasingly making their way to the Oval Office in the White House, prompting President Jimmy Carter to issue an executive order in 1977. This order directed CEQ to issue formal NEPA regulations. Although CEQ had adopted guidelines under President Nixon, this guidance focused primarily on implementing the EIS process. Under President Carter's administration, CEQ's authority was strengthened. The earlier guidelines would be replaced with binding regulations that expanded their scope to cover the entire NEPA process, rather than simplify the EIS process. These regulations were to provide specific direction for reducing paperwork and delay and for focusing the NEPA process on analyzing only important environmental issues.[7]

Eight years after NEPA was enacted, CEQ promulgated its formal NEPA implementing regulations (40 CFR 1500-1508) in November 1978. CEQ carefully crafted language to address concerns that had emerged during the first several years of experience with the act. To promote efficiency, the regulations provided direction for restricting the length of EISs, encouraged agencies to set time limits, and stressed the need to focus on important issues and to de-emphasize insignificant ones. The regulations called for other streamlining methods, such as reliance on "scoping" to help agencies focus on key issues, incorporating material by reference, and combining EISs.

The EA, the subject of this book, is one of several outgrowths of this reform. The statutory language of NEPA does not specifically mention the EA or any document meeting the description of what we now refer to as the EA. Instead, the NEPA implementing regulations introduce the EA as a means to streamline NEPA compliance by allowing agencies to prepare a concise document demonstrating that a proposed action is incapable of significant environmental impacts, in lieu of the "detailed statement." The regulations refer to the EA as a "concise document," indicating a clear intent that the EA be substantially shorter and less expensive to prepare than an EIS. However, as will be explained in subsequent chapters of this book, the EA—while indeed usually shorter and cheaper than a typical EIS—also serves a distinctly different purpose than an EIS. While EISs serve as the "detailed statement" of environmental impacts from actions "significantly affecting the human environment," EAs serve to demonstrate that actions lack significant environmental impacts. Understanding this difference is the core message of this book.

1.3.6 Public comment review process

CEQ's draft regulations underwent extensive public hearings and comment review. CEQ met with every agency in the federal government and solicited the views of nearly 12,000 organizations, private individuals, and state and local agencies. Chambers of commerce, the Building and Construction Trades Department of the AFL-CIO, and the National Resources Defense Counsel, to name just a few, were among those consulted. Nearly 500 written comments were received on the draft regulations, the majority of which were favorable and expressed support for the regulations.[8] The overwhelming consensus was that NEPA benefited the public. Favorable responses notwithstanding, many comments on the draft regulations complained that the process as experienced to date was too lengthy, resulted in large quantities of needless detail, and needed to be streamlined. A total of 340 amendments were made to the draft regulations before they were promulgated in 1978 as final regulations.[9] Case law was codified into the regulations, increasing the likelihood that the courts would defer in their favor. Since their promulgation, only one clause in the regulations (§1502.22) has been amended: a requirement to perform a "worst case" analysis in circumstances involving incomplete or unavailable information.

1.3.7 NEPA today

There is little doubt that a principal driver of NEPA compliance has been the accumulation of NEPA litigation communicating clearly that NEPA is to be taken seriously. Perhaps a more important driver of change has been a generally improved environmental ethic within most federal agencies and a top-down commitment to NEPA compliance. Many agencies have created environmental compliance offices and have recruited a new generation of trained environmental specialists to deal with NEPA and other environmental requirements. Most agencies are now making a good faith effort to incorporate NEPA's intent into their decision making process. This is illustrated by a statement made by former Secretary of Energy Watkins, who testified thus before Congress:

> As Secretary of Energy I quickly learned that the NEPA process was not being used to provide complete and unbiased information that top-level managers needed to make the best decisions. Therefore, I established new policies to enhance and reinvigorate the DOE NEPA process.[10]

Senator John Chafee testified:

> I believe, Mr. President, when historians look back to the years 1969 and 1970, they will say those were watershed years in terms of the U.S. environmental movement. ... Of all these and other significant actions that took place in those two years, few can rival in importance the creation of the National Environmental Policy Act. ... NEPA has been a tremendous success and has changed forever the way our government makes decisions affecting the environment.[11]

Until the passage of NEPA, federal decisions were made without the benefit of environmental information. Under NEPA, environmental considerations in decision making are much better integrated with economic and technical factors. Although there have been many problems with the NEPA process, including an abundance of litigation leading to sometimes long and convoluted documentation, NEPA's regulations have proved remarkably stable and durable. Agencies are becoming more adept at discovering adverse effects and modifying their proposed actions in the early planning process so as to avoid such impacts. As the authors of NEPA hoped, gross environmental impacts are becoming less frequent as managers gain access to information in the earliest stages of project planning for use in making the wisest possible decisions. Many problems in NEPA implementation have been reduced to issues of interpretation of the regulations, the need for agencies to implement effective procedures and methodology, and the need for qualified environmental specialists.

It must also be appreciated that NEPA is unique among federal environmental laws for its comprehensive nature. Most environmental statutes address a single environmental resource, as evidenced in names such as the Clean Air Act, Clean Water Act, Endangered Species Act, and National Historic Preservation Act. NEPA requires consideration of impacts to air, water, endangered species, and historic preservation, but also requires consideration of other environmental resources and impacts not specifically spelled out in the name or content of the statute. Proponents and practitioners of these other narrowly scoped acts can easily fall into a trap of only narrowly considering one resource at the possible expense of another. However, NEPA requires a comprehensive balancing of the interests of all environmental resources, together with nonenvironmental considerations.

1.3.8 *NEPA's effect around the world*

Few environmental statutes have contributed more to the long-term preservation of environmental quality. NEPA's importance is derived not from any substantive requirement to protect the environment, but from the effect it has had in shaping and influencing the federal planning apparatus from the outset. Perhaps the most important factor is that NEPA sets forth an environment-planning process with the expressed purpose of forcing agencies to look before they leap. This difference makes NEPA a proactive environmental initiative rather than a reactive one. In opening the early federal planning process to public review and debate, federal agencies have been compelled to take environmental considerations into account in the "light of day."

Although far from perfect, the process for determining the threshold level of significance provides an environmental system of checks and balances. Within the public forum, compromises are struck as competing environmental and nonenvironmental interests vie for influence. Slowly, this democratic process appears to be evolving toward a middle ground where environmental interests are balanced against society's need to develop and prosper.

1.3.9 *NEPA's global precedent*

NEPA was enacted to "promote efforts which will prevent or eliminate damage to the environment and biosphere and stimulate the health and welfare of man."[12] Beyond American shores, NEPA has established a global precedent that has been emulated by scores of other nations. The influence of NEPA has permeated virtually every corner of the globe and has the distinction of being one of the most copied acts in the world. Today

NEPA has been emulated, in one form another, by more than 25 states and more than 100 countries worldwide. The Organization for Economic Cooperation commended the United States for its "exemplary practices" in environmental impact analysis and public participation.[13] The European Economic Community now requires its members to comply with an environmental process similar to that of NEPA. Moreover, the World Bank requires funding recipients to prepare an NEPA-like analysis to evaluate environmental consequences of their projects. To fully appreciate the significance of this act, we must understand the historical context in which it was enacted. The next sections of this chapter provide the historical backdrop, an overview of the act, and its subsequent regulations.

1.4 Brief overview of NEPA

First and foremost, NEPA is a statement of our national will to protect the environment. Less a regulatory statute than a policy act, NEPA establishes a fundamental principle by which the federal government is to conduct its operations. On the surface, NEPA appears to be a weak law: Its broad scope challenges an agency's capability to handle a broad array of complex and diverse issues; it sets no substantive environmental standards and defines no enforcement mechanisms beyond an agency's discretion.

In practice, however, NEPA is a balance of competing interests, whereby perceived weaknesses are tempered with complementary strengths. The genius of its initiators lay in their vision of the act's power in the face of such weaknesses. The lack of substantive standards provides planners with a great degree of flexibility in planning actions and is more than compensated for by a plethora of such standards in other environmental laws that can be identified and integrated within NEPA. The lack of an enforcement mechanism at first created a vacuum, but this has been amply filled by the courts as parties have challenged agency actions under NEPA's provisions.

1.4.1 Procedure for passing laws and regulations

Before addressing the act and its subsequent regulations, a brief explanation of the process followed in the United States for promulgating laws and regulatory requirements is instructive. Statutes are enacted as public laws by Congress and signed by the president. Public laws are collected by the Office of the *Federal Register* and printed as a volume titled *The United States Statutes at Large*. Public laws are codified into the entire body of federal laws in the *U.S. Code* (USC), which is a collection of all United States laws. It should be noted that section numbers used in the public law do not necessarily follow the same section numbering used in the USC.

Many laws enacted by Congress require federal agencies to issue implementing regulations. Typically, an agency first prepares a draft regulation, which is published as a proposed rule in the *Federal Register* (FR). The public is invited to submit comments within a specified period (usually a minimum 90 days). Public comments are considered and incorporated into the final regulation, which is published in the *Federal Register*. After publication, regulations are codified annually in the *Code of Federal Regulations* (CFR). This public involvement process is remarkably similar to the public involvement process used in the review of draft EISs under NEPA. The CFR is divided into 50 titles representing broad subject areas; Title 40 concerns the environment. The CFR is referenced by title and part number (e.g., 40 CFR parts 1500-1508). A section in the CFR is analogous to a paragraph in a text and normally designates a single concept (e.g., 1500.1—Purpose).

1.4.2 Regulatory nomenclature

Before introducing the details of NEPA the nomenclature used in referring to NEPA and its regulations needs to be introduced. Throughout the following discussion, the term "National Environmental Policy Act" is shortened to "NEPA" or "act." Similarly, the term "CEQ NEPA regulations" is abbreviated to simply "regulations." For brevity, references to a particular section of the CEQ regulations (40 *Code of Federal Regulations* [CFR] 1500-1508) are abbreviated so that they simply cite the specific section number in the regulations where the provision can be found. For example, a reference such as "40 CFR 1500.1" is shortened to the more convenient expression "§1500.1."

1.4.3 Overview of NEPA's mandate

NEPA, the statute, contains just three sections (a statement of purpose followed by two titles) and is only about five pages long. Despite the brevity of the statute, NEPA has had a profound effect on the federal decision making process. Title I declares a national environmental policy and sets forth procedural requirements that must be followed in pursuing proposed actions, and Title II creates the Council on Environmental Quality (CEQ).

NEPA is most notable for three principal elements that it brings to bear in achieving environmental protection: (1) declaration of the world's first national policy on the environment; (2) establishment of an "action-forcing" mechanism for implementing this policy; and (3) creation of a council for implementing NEPA and elevating environmental concerns directly to the presidential level. Although its components are briefly described in the following paragraphs, the reader is encouraged to read the act in its entirely (see Appendix A). A summation of the act follows.

1.4.4 Purpose of NEPA

The purpose of NEPA is to establish a national environmental policy and a new federal organization—the Council for Environmental Quality. Specifically, the purpose of the act is to

> ... declare a national policy which will encourage productive and enjoyable harmony between man and his environment; promote efforts which will prevent or eliminate damage to the environment and biosphere, and stimulate the health and welfare of man; enrich the understanding of the ecological systems and natural resources important to the Nation; and to establish a Council on Environmental Quality.

1.4.5 Title I—declaration of the National Environmental Policy Act

This title is the heart of NEPA, as it not only announces a national environmental policy and goals, but also creates specific responsibilities for federal agencies. Section 101 defines the nation's environmental policy and is sometimes referred to as the "spirit of the law." Section 102 provides the procedural or action-forcing mechanism for carrying out the policy established in Section 101. Often referred to as the "letter of the law," Section 102 specifies the procedural requirements regarding preparation of an EIS.

Section 101 declares a policy to use all practical means, including financial and technical, to promote the general well-being of the environment. Specifically, all practicable means will be used to:

1. Fulfill the responsibilities of each generation as trustee of the environment for succeeding generations.
2. Assure for all Americans safe, healthful, productive, aesthetically, and culturally pleasing surroundings.
3. Attain the widest range of beneficial uses of the environment without degradation, risk to health or safety, or other undesirable or unintended consequences.
4. Preserve important historic, cultural, and natural aspects of our national heritage and maintain, wherever possible, an environment that supports diversity and variety of individual choice.
5. Achieve a balance between population and resource use that will permit high standards of living and a wide sharing of life's amenities.

6. Enhance the quality of renewable resources and approach the maximum attainable recycling of depletable resources.

Court decisions have held that the goals set forth in Title I impose only a procedural duty on federal agencies to consider NEPA's goals when making a final decision regarding proposed actions. Thus federal agencies are bound to comply with the procedural requirements of NEPA but are not obligated to make decisions based on preserving environmental quality. Just because the NEPA process identifies alternatives or modifications to actions that could reduce environmental impacts, agencies are not necessarily obligated to pursue those alternatives or modifications. Still, the ultimate accomplishment of NEPA is when the process identifies ways to reduce environmental impacts whose value is not overridden by economics or other factors related to an agency's mission. "The phrase look before you leap" does not necessarily imply that an agency may not indeed ultimately leap; it just means that the agency will be armed with environmental information prior to leaping. NEPA was intended to reduce the probability that agencies make mistakes resulting in unnecessary environmental impacts; it was never intended to prevent environmental impacts.

The phrase "life's amenities," mentioned in item 5 in the preceding list, requires special explanation. In respect to the environment, amenities can be thought of as those aspects and resources that are considered to be attractive, pleasurable, or to possess desirable attributes. For example, an environmental amenity may be a natural, unobstructed scenic area that the general public regards as beautiful and peaceful. NEPA does not limit its coverage only to those environmental resources such as air and water whose protection has distinct economic or safety implications; its coverage also extends to more esoteric issues such as aesthetics and cultural features. The breadth of NEPA's scope is perhaps what distinguishes NEPA most from the other more narrowly focused federal environmental statutes.

Section 102 requires that all federal agencies shall, "to the fullest extent possible," interpret and carry out their policies and duties in accordance with NEPA. Among the key provisions that have had far-reaching effect are the following citations directing all federal agencies to:

> ... utilize a systematic, interdisciplinary approach which will ensure the integrated use of the natural and social sciences and the environmental design arts in planning and in decision making which may have an impact on man's environment (Sec. 102[2][A])

and

> ... include in every recommendation or report on proposals for legislation and other major federal actions significantly affecting the quality of the human environment, a detailed statement by the responsible official—on (i) the environmental impact of the proposed action (Sec. 102[2][C]).

The first of these provisions challenges agencies' creativity and innovation by requiring them to go beyond their traditional mission-oriented thinking when planning new actions. The latter provision is at the core of NEPA's enforcement—as the regulations have defined specific requirements for the "detailed statement" and the courts have further defined and clarified the requirements of this provision through case law. The act provides surprisingly little detail regarding the specific scope and content of the EIS, and does not even mention anything even remotely resembling the EA. This task was left to the drafters of the regulations. While the concept of the EIS can be traced to the statutory language of NEPA, the EA is strictly a product of the NEPA implementing regulations. Indeed, the EA can be considered a response to one of the unintended realities encountered once NEPA was implemented—that the term "significantly" is one of judgment not measurement, and that preparing "detailed statements" for each and every federal action capable of "significantly" affecting the environment in the eyes of some interested party would place an untenable logistical burden on agencies whose cost would not be justified by the ensuing value to the environment.

1.4.6 Title II—Council on Environmental Quality

It was widely understood that a national policy could not be self-implementing without some independent oversight. Congress recognized that a council of some sort would be needed to oversee the act and ensure its implementation.

Accordingly, Title II establishes the CEQ, or "council," as it is referred to in the regulations. The council reports directly to the president, who has ultimate authority for ensuring that agencies conduct their actions in a manner consistent with the national policy. Title II defines eight specific "duties and functions," which include advising the president, resolving interagency disagreements, gathering information about the quality of the environment, and making recommendations in respect to policy and legislation.

Patterned after the Council on Economic Advisors, this council advises the president on environmental matters. The president, with the Senate's concurrence, appoints three members to the council. As part of the Executive Office of the President, the council reports directly to the president; one member is appointed by the president to act as chairman.

It is noteworthy that the act delegates no enforcement powers to CEQ in respect to the implementation of Title I. As mentioned earlier, an executive order was issued granting CEQ authority for issuing guidelines for implementing Title I. CEQ was not empowered to create regulations until several years after NEPA's enactment. However, with one exception, even the regulations do not empower CEQ to actually oversee or enforce agency compliance with Title I. This exception involves resolving disagreements between federal agencies "concerning major federal actions that might cause unsatisfactory environmental effects."

Interestingly, the EPA was established in 1970, the first year during which NEPA was implemented. Although the product of the same environmental movement that led to NEPA, EPA was not a direct product of NEPA. The mission of EPA was substantially greater than CEQ. EPA was established as a regulatory agency to implement and enforce requirements, and issue permits under multiple resource-specific environmental protection acts. Its scope ultimately encompassed the Clean Air Act; Clean Water Act; Resource Conservation and Recovery Act (RCRA); Comprehensive Environmental Response, Compensation, and Liability Act (Superfund); and others. EPA also came to play a key role in reviewing EISs and certain other roles in carrying out NEPA. Nevertheless, the office specifically established by NEPA, CEQ, continues to function in its role as an advisory group to the president on environmental issues. When issues or questions arise in the implementation of NEPA that cannot be settled by agencies or by EPA, they are commonly elevated to CEQ for direction and guidance. Of course, even CEQ is not a court; many disputes in NEPA are ultimately decided in the courts.

1.5 NEPA regulations

The CEQ Regulations have nine parts, summarized in Table 1.1. A copy of the regulations is provided in Appendix B. As discussed previously, compliance is defined primarily as a matter of procedure rather than as a substantive environmental standard. For this reason, the regulations are focused on how to implement Title I, Section 102(2), of the act. Among the items covered in the regulations is the definition and use of the EA.

The deference given to these regulatory requirements has been reinforced by countless court cases. Nevertheless, the regulations have been crafted so as to allow ample opportunity for both innovative implementation methods as well as technical disagreements on how they should be implemented. Moreover, a review of case law on certain issues sometimes reveals differing interpretations among the courts, or what appear to be very subtle distinctions between findings.

As noted earlier, most environmental regulations are highly prescriptive and are assigned clear authority (mostly to the EPA) for their enforcement. The regulations often define strict thresholds of compliance and even prescribe technical means such as specific models or laboratory tests for determining compliance. NEPA's regulations are clear on a number of procedural matters, but a large measure of compliance is left to the discretion of individual agencies, subject only to court authority when challenged.

This means that NEPA compliance can be both flexible and variable, responding to such factors as the presence or absence of project opposition and the culture of the agency. The regulations provide an essential but incomplete guide to successful NEPA implementation. This situation is particularly challenging to the NEPA practitioner. In addition to learning the regulations, the practitioner should be aware of key court findings and of available scientific and technical methods for characterizing environmental impacts. Additional direction for complying with NEPA appears in other laws, regulations, and executive orders. Moreover, there is a body of official guidance issued by the CEQ that does not have the force of regulation but is widely followed and sometimes cited in court. Key among this nonregulatory guidance are the "Forty Most Asked Questions" on NEPA, which address in question and answer format many practical elements encountered in NEPA practice. Several of the forty questions address the preparation and use of EAs and constitute the best "official" guidance from CEQ specifically geared to the EA.

1.6 Overview of the NEPA planning process

This section provides a general overview of the NEPA process with emphasis on describing the three levels of NEPA compliance. A detailed accounting of every aspect and intricacy inherent to the NEPA process is beyond the scope of this text. For a detailed review, the reader is referred to the companion text, *NEPA and Environmental Planning*.[5]

Details governing the actual implementation of NEPA can vary, particularly with respect to the way individual agencies choose to implement specific aspects of their respective processes. For requirements governing specific circumstances, the reader is referred to the regulations, and the agency's internal orders and NEPA implementation procedures.

1.6.1 When must NEPA begin?

The EIS process must be started so that it "can be completed in time for the final statement to be included in any recommendation or report on the proposal" (§1502.5); thus the NEPA process is to begin as close as possible to the time in which an agency is developing or is presented with a proposal. This provision is interpreted to apply to preparation of Environmental Assessments (EAs) as well as an EIS, as all federal actions are considered potentially subject to the requirements of an EIS until it can be demonstrated otherwise.

1.6.2 When must NEPA be completed?

The NEPA process must be initiated early enough so that it can contribute to the decision making process and will not be used to rationalize or justify decisions already made. This timing requirement is met when a NEPA analysis has been prepared in time to meet the decision deadline, but not so early that it cannot meaningfully contribute to the decision making process (§1502.5).

1.6.3 Three levels of NEPA compliance

The NEPA process can be viewed as consisting of three levels of planning and environmental compliance. These three levels, defined from the least to the most demanding, are:

- Categorical Exclusion (CATX)—Some federal actions might qualify for a CATX, thus excluding them from further NEPA review and documentation requirements.
- Environmental Assessment (EA)—If a federal action does not qualify for a CATX, an EA may be prepared to determine whether a federal action qualifies for a Finding of No Significant Impact (FONSI), thus exempting it from the requirement to prepare an EIS.
- Environmental Impact Statement (EIS)—In general, an EIS must be prepared for proposed federal actions that do not qualify for either a CATX or a FONSI.

1.6.4 Initiating the NEPA process

Care must be exercised in selecting a manager who can exercise good judgment and who possesses experience in preparing NEPA analyses. Inexperienced management or analysts have been one of the principal causes responsible for cost overruns, poor planning, and flawed analysis.

In one instance, an NEPA technical support document was prepared for a waste management project at a federal installation operated by the Department of Energy. The group manager lacked experience preparing environmental impact studies and shied away from making key decisions. When decisions were made, they were frequently formulated with little regard to future ramifications. While this manager eventually left her position, the quality of this technical planning process and other similar environmental compliance efforts was questionable. If management lacks experience, it is vital that the counsel of experienced practitioners be sought.

Examining proposals for existing NEPA coverage: The NEPA planning process normally begins when a need for taking action has been identified (see oval-shaped rectangle in the upper left hand corner of Figure 1.1). Figure 1.1 provides a simplified overview of the entire NEPA process (including preparation of the EIS).

A new action might fall within the scope of an existing EIS or EA. If the proposal is sufficiently covered under existing NEPA documentation, the agency can proceed with the action (with respect to NEPA's requirements). Existing NEPA documentation should be examined to determine whether the proposal has been subject to a previous NEPA review (first and second decision diamond, Figure 1.1). Reaching such a determination often requires exercising a substantial degree of professional judgment. Under certain conditions, the proposal might require supplementing an existing EIS (§1502.9[c]).

Categorically excluding actions: A proposal might be excluded from further NEPA review if it falls within an existing class of actions (i.e., CATXs) that have been previously determined to result in no significant impact (cumulative or otherwise) and for which preparation of an EA/EIS is therefore not required (see third decision diamond, Figure 1.1). Each agency is required to prepare a list of CATXs as part of its individual NEPA implementation procedures. If a CATX is applicable, the agency can proceed with the action (with respect to NEPA requirements).

It is important to note that, under §1508.4, agencies are also required to "... provide for extraordinary circumstances in which a normally excluded action may have significant environmental impacts." For more information on the application of CATXs, the reader is referred to *The NEPA Planning Process*.[6]

If the proposal has not been excluded, the agency should review its NEPA implementation procedures for guidance in determining the appropriate level of NEPA compliance.

CATXs and Case Law. Care must be exercised in applying a CATX. An agency's review process and administrative record should support its decision to apply the CATX. Witness a recent case, in which the Department of Interior was sued for allowing a company to bioprospect for microbes in Yellowstone National Park. The Department of Interior argued that the bioprospecting activities fell under its CATX for "day-to-day resource management and research activities." According to the court, it is inappropriate to claim that an action is covered by a CATX when the agency lacks "evidence in the administrative record or elsewhere that such a determination was made at the appropriate time." The court went on to state that use of this CATX was questionable because (1) the commercial exploitation of natural resources is probably not

24   *Preparing NEPA Environmental Assessments*

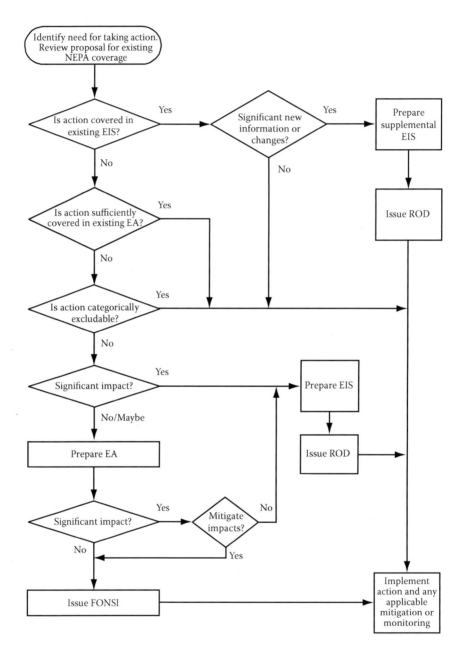

Figure 1.1 Overview of typical NEPA process.

Chapter one: Overview and historical development of NEPA

equivalent to "day-to-day resource management and research activities"; and (2) such activities involved extraordinary circumstances associated with "unique geographic characteristics" and "ecologically significant or critical areas," thus making the proposed action ineligible for a CATX under the agency's own rules.[7]

1.6.5 Environmental assessment

If the environmental review determines that the action is not eligible for a CATX, the agency can choose to prepare an EA to determine whether the proposal could significantly affect quality of the human environment (see box labeled "Prepare EA," Figure 1.1).

If the agency believes that the proposal would result in a significant impact, it can choose to prepare an EIS without first writing an EA (see fourth decision diamond labeled "Significant impact?" Figure 1.1). To facilitate future analysis, an agency can also prepare an EA even if it plans to eventually prepare an EIS for the action.

With respect to NEPA's requirements, the agency is free to pursue any course of action as long as it does not result in a significant impact. A Finding of No Significant Impact (FONSI) is prepared if, based on the EA,

---

WHEN AN ENVIRONMENTAL ASSESSMENT SHOULD BE PREPARED

"Agencies shall prepare an environmental assessment ... when necessary under the procedures adopted ... to supplement these regulations. ... An assessment is not necessary if the agency has decided to prepare an environmental impact statement."

(§1501.3[a])

"Agencies may prepare an environmental assessment on any action at any time in order to assist agency planning and decision making."

(§1501.3[b])

"Briefly provide sufficient evidence and analysis for determining whether to prepare an environmental impact statement or a finding of no significant impact."

(§1508.9[a][1])

the decision maker concludes that no significant impact would occur (see box labeled "Issue FONSI?" Figure 1.1). If the EA leads to a FONSI, an EIS is not required, and the agency can proceed with the action (with respect to NEPA's requirements).

Once a FONSI has been issued, the agency is generally free to proceed with the action (with respect to NEPA's requirements), in accordance with any applicable mitigation or monitoring measures. This process is described in detail in Chapter 5.

When should an assessment be prepared?—An EA may be prepared on any action at any time to foster agency planning objectives, or to provide sufficient evidence for determining whether an EIS is required (§1501.3[b]).

Three purposes of an assessment—Preparation of an EA can serve three basic purposes (§1508.9[a]). The most important of these is to briefly provide sufficient evidence and analysis for determining whether to prepare an EIS or a FONSI.

The assessment and significant impacts: Two options exist if the proposal could result in a significant impact (see fifth decision diamond labeled "Significant impact," Figure 1.1), (1) mitigate the potentially significant impacts to the point of nonsignificance (see box labeled "Mitigate impacts?" Figure 1.1); or (2) prepare an EIS.

NEPA'S REQUIREMENTS FOR PREPARING AN EIS

... include in every recommendation or report on proposals for legislation and other major federal actions significantly affecting the quality of the human environment, a detailed statement by the responsible official ... on

  i. The environmental impact of the proposed action
 ii. Any adverse environmental effects which cannot be avoided should the proposal be implemented
iii. Alternatives to the proposed action
 iv. The relationship between local short-term uses of man's environment and the maintenance and enhancement of long-term productivity
  v. Any irreversible and irretrievable commitments of resources which would be involved in the proposed action should it be implemented

(Section 102[2][c] of NEPA)

1.6.6 Environmental impact statement

An EIS is prepared to investigate environmental consequences and alternatives for pursuing a proposal. An EIS must be prepared (with few exceptions) if a federal proposal cannot be categorically excluded and is not eligible for a FONSI (see box labeled "Prepare EIS" Figure 1.1).

The EIS must be used by decision makers in reaching a final decision regarding the course of action to be taken (§1502.1). A Record of Decision (ROD) is issued, publicly announcing the course of action that the agency has chosen to pursue. On completing this process, the agency is free to pursue the course of action described in the ROD. Once an ROD has been issued, the agency is free to proceed with the action (with respect to NEPA's requirements), in accordance with any applicable mitigation or monitoring measures. The reader is referred to the companion text, Environmental Impact Statements, for a detailed discussion of EISs.[8]

Supplementing an EIS as indicated in Figure 1.1 (see diamond labeled "Significant new information or changes"), under certain circumstances, an agency must supplement a draft or final EIS (§1502.9[c]). Once the supplemental EIS process has been completed, the agency issues an ROD documenting its decision with respect to the course of action that will be taken. The agency is then free to proceed with its decision, in accordance with any applicable mitigation or monitoring measures.

References

1. O'Neil, D., "Project Chariot: How Alaska Escaped Nuclear Excavation," *The Bulletin of the Atomic Scientist* (December 1989): 35.
2. Caldwell, L.K., "A Constitutional Law for the Environment: 20 Years with NEPA Indicates the Heed," *Environments*, No. 10 (December 1989).
3. Senate Report No. 91–296,
4. 15 U.S.C. §§1021–1025.
5. Eccleston, C., *NEPA and Environmental Planning Tools, Techniques, and Approaches for Practitioners*. CRC Press 2008.
6. Executive Order 11514, March 5, 1970, Protection and Enhancement of Environment Quality, as amended by Executive Order 11991, May 24, 1977.
7. CEQ, Preamble to Final CEQ NEPA Regulations, 43 Fed. Reg. 55978, Section 1, November 29, 1978.
8. CEQ, Preamble to Final CEQ NEPA Regulations, 43 Fed. Reg. 55978, Section 3, November 29, 1978.
9. CEQ, Preamble to Final CEQ NEPA Regulations, 43 Fed. Reg. 55978, Section 6, November 29, 1978.
10. Watkins, J., former secretary, U.S. Department of Energy (Admiral, U.S. Navy [ret.]), testimony before the House Armed Services Committee, 1992.
11. Congressional Record, p. S4141, Senate Proceedings, March 25, 1992.

12. NEPA Section 2, 42 U.S.C. § 4321.
13. CEQ, *The National Environmental Policy Act: A Study of Its Effectiveness After Twenty-Five Years*, 1997.

chapter two

General concepts and requirements

> I have been complimented many times and they always embarrass me; I always feel that they have not said enough.
>
> —Mark Twain

2.1 Introduction

The National Environmental Policy Act (NEPA) regulations (regulations) developed by the Council on Environmental Quality (CEQ) in 40 CFR 1508 et seq. were primarily written with the objective of specifying regulatory requirements governing preparation of Environmental Impact Statements (EISs). The focus was clearly on preparing EISs. Given this focus, only limited attention was devoted to describing regulatory requirements that also apply to the preparation of Environmental Assessments (EAs), the subject of this book. In fact, the concept of an EA appears to have been more an afterthought than a thoroughly defined procedural mechanism. Hence a vacuum has existed with respect to determining which requirements specified in the regulations also apply to preparation of EAs. One of the objectives of this book is to bridge this gap.

Accordingly, this chapter identifies and describes those regulatory provisions that are interpreted to apply to the preparation of EAs as well as EISs. A thorough understanding of these basic requirements and concepts is essential in effectively planning agency actions and in complying with NEPA's requirements when preparing EAs, and sets the stage for application of these principles in subsequent chapters. For a detailed discussion of NEPA's essential concepts and efficiency requirements, the reader is referred to Chapters 3 and 4 of *NEPA and Environmental Planning*.[1]

The following sections contain basic requirements that are interpreted to apply to the entire NEPA process, including preparation of EAs. The reader is encouraged to read the actual regulatory provisions that are cited.

2.2 NEPA is a planning and decision making process

NEPA is often incorrectly viewed as a document preparation process. While preparation of environmental documents is an integral and necessary component, it is not why NEPA was enacted. In fact, one of the original purposes for CEQ promoting the EA in the regulations was to provide a mechanism for meeting NEPA's environmental planning objectives without necessarily having to prepare a large document. The central purpose of NEPA is to provide decision makers with information on which to base decisions. This includes the decision whether to prepare an EIS or proceed with an action with the understanding that no potentially significant impacts could result. Table 2.1 announces the purpose of NEPA.

2.2.1 Reasonable alternatives

Traditionally, the primary reason for preparing an EA has been to determine whether a proposed action would significantly impact environmental quality or could qualify for a finding of no significant impact (FONSI). Yet, an EA can also be used as a tool for planning actions and determining an optimum course of action. For example, in one recent case, an agency had already made a decision to take some kind of action to avoid an environmental fine. An EA was prepared to evaluate "all reasonable alternatives." The agency reported that this EA provided an effective planning tool for specifically determining the best course of action to take.[2]

2.3 Interim actions

Periodically, an agency might need to implement individual project or program element actions that fall within the scope of an ongoing NEPA analysis. As depicted in Table 2.2, pursuing an action under such circumstances would normally constitute a violation of NEPA's regulatory provisions. Actions that might legitimately proceed in advance of completing an ongoing NEPA review process are referred to as interim actions. Table 2.2

Table 2.1 NEPA's purpose

Ultimately, of course, it is not better documents but better decisions that count. NEPA's purpose is not to generate paperwork—even excellent paperwork—but to foster excellent action. The NEPA process is intended to help public officials make decisions that are based on understanding of environmental consequences, and take actions that protect, restore, and enhance the environment.
(1500.1[c])

Table 2.2 Requirements for proceeding with an action in advance of completing the EIS process

Non-programmatic EIS

Until an agency issues a ROD, no action concerning the proposal shall be taken that would

1. Have an adverse environmental impact,
2. Limit the choice of reasonable alternatives

(§1506.1[a])

Programmatic EIS

While work on a required program environmental impact statement is in progress and the action is not covered by an existing program statement, agencies shall not undertake in the interim any major federal action covered by the program that might significantly affect the quality of the human environment unless such action

1. Is justified independently of the program
2. Is itself accompanied by an adequate environmental impact statement
3. Will not prejudice the ultimate decision on the program. (An interim action prejudices the ultimate decision on the program when it tends to determine subsequent development or limit alternatives.)

(§1506.1[c])

summarizes important interim action provisions in the regulations. Chapter 10 of Eccleston's book *NEPA and Environmental Planning* provides an in-depth discussion of the requirements governing interim actions.[3]

2.3.1 Eligibility for interim action status

Section 1506.1 of the regulations places specific limitations and requirements on actions permitted to take place prior to completing NEPA. As depicted in Table 2.2, §1506.1 separates interim action requirements into two categories: Non-programmatic EIS and Programmatic. In the context of NEPA, a "program" consists of a coordinated sequence of related actions.

Before an interim action can be implemented, §1506.1[c][2] requires that the proposal must first have been adequately investigated in an EIS. Strictly interpreted, this requirement leads to a paradox with potential repercussions in terms of cost, schedules, and resource requirements because it appears to preclude actions that qualify for a categorical exclusion (CATX) or finding of no significant impact (FONSI), but have not been the subject of an EIS. Thus even if an interim action can be shown to have no significant impact, a strict interpretation leads to the unreasonable conclusion that it would still have to be reviewed in an EIS before the agency could pursue the action. The text *NEPA and Environmental Planning* provides a mechanism for resolving this paradox.[4]

2.4 Integrating NEPA with other requirements

Agencies are instructed to integrate NEPA with other environmental planning and review efforts (e.g., regulatory requirements, permits, agreements, studies, project planning) so that procedures run concurrently rather than consecutively. Such practice can:

1. Avoid duplication of effort.
2. Avoid unnecessary research, analyses, and writing.
3. Avoid possible project resentment among other agencies and the public.
4. Reduce project delays.
5. Minimize environmental compliance costs.
6. Result in more effective decision making.

These objectives apply both to the EIS and EA processes, although they may be even more pressing for actions addressed in the shorter EA process. Table 2.3 summarizes some of the pertinent regulatory citations that provide direction for integrating NEPA with other processes. Where possible, agencies are instructed to cooperate with state and local agencies to eliminate duplication.

2.4.1 Integrating environmental design arts

Under NEPA, agencies are required to "integrate environmental design arts into their planning and in decision making. ..."[5] This requirement is interpreted to mean that disciplines such as architecture and urban planning (the environmental design arts) must be integrated together with the natural and social sciences into the agency's planning process so that federal actions are blended more naturally into the environment. Compliance with this requirement implies that an EA should be used to assist the agency in planning actions as well as determining whether an EIS is required.

2.5 Conducting an early and open process

Agencies are required to conduct an early and open process. NEPA documents are required to be prepared and publicly issued at the same time as other planning documents (§1501.2[b]). Fulfilling this requirement is crucial if an agency is to truly use NEPA as a planning and decision

Table 2.3 Integrating NEPA

Integrating the NEPA process into *early planning*
(§1500.5[a], §1501.1[a], emphasis added)

Integrate the requirements of NEPA with *other planning* and environmental review procedures ... so that all such procedures run concurrently rather than consecutively
(§1500.2[c], emphasis added)

Agencies shall *integrate* the NEPA process with *other planning* at the *earliest possible time* ...
(§1501.2, emphasis added)

Identify *other environmental review* and *consultation requirements* ... prepare other required *analyses* and *studies* concurrently with, and *integrated* with ...
(§1501.7[a][6], emphasis added)

Any environmental document in compliance with NEPA may be *combined* with any other agency document ...
(§1506.4, emphasis added)

Agencies shall *cooperate* with state and local agencies to the fullest extent possible to reduce duplication between NEPA and comparable state and local requirements ...
(§1506.2[c], emphasis added)

---

making tool, which informs decision makers and the public about the consequences of potential actions. The need for an early and open process applies to the EA as well as the EIS process.

2.6 Public involvement

In the past, some agencies have operated under the mistaken belief that the EA process does not require public involvement. Many EAs were written, used as the basis to proceed with an action, and buried in a file cabinet. However, an EA is, in fact, a "public document" (§1508.9). Although rarely as conspicuously publicized as most EISs, EAs are still part of an agency's public record and must be made accessible to the public. The increased use of the World Wide Web has made EAs even more accessible to interested members of the public. Moreover, agencies are instructed to make diligent efforts to involve the public in preparing and executing their NEPA implementation procedures (§1506.6[b]).[6] This direction has been recently amplified by the Obama administration's promotion of greater transparency throughout the federal government.

Agencies are required to "provide public notice of NEPA-related *hearings, public meetings*, and the *availability of environmental documents* so as to inform those people and agencies who may be interested or affected (§1506.6[a], emphasis added)." The term "environmental documents" is defined to include EAs (§1508.10). Note that the term "hearing" does not necessarily refer to the highly formalized hearings conducted outside of the context of NEPA by some agencies.

2.7 Scoping

Agencies are required to perform an "early and open" process in determining the scope of an EIS. However, the regulations are silent with respect to the application of scoping during the EA process. Nevertheless, application of scoping (whether internal or public) during the EA process is crucial to the objective of providing relevant information to the decision maker. Chapter 11 of the text *NEPA and Environmental Planning* provides a detailed discussion of NEPA's concept of scope.[7] The purpose of scoping is to promote efficiency by focusing a NEPA document on issues that are truly relevant or otherwise of concern to the public. The efficiencies that can be gained through scoping therefore enhance and complement the efficiencies that can be gained by preparing an EA rather than EIS to address actions lacking potentially significant environmental impacts.

2.8 Systematic and interdisciplinary planning

Agencies are required to use a systematic and interdisciplinary approach in implementing their NEPA process. This requirement has been interpreted to extend to the preparation of EAs as well as to EISs.[8]

2.8.1 Systematic

The term systematic places a mandate on agencies to utilize a logical, ordered, and methodological approach in which each stage of the EA process builds upon previous stages.

2.8.2 Interdisciplinary

The requirement to perform an interdisciplinary approach places a burden on agencies to ensure that the analysis is performed by knowledgeable specialists who possess expertise in the disciplines for which they have been assigned responsibility.

The terms multidisciplinary and interdisciplinary are not equivalent. A multidisciplinary approach refers to a process in which specialists perform their assigned tasks with little or no interaction. In contrast, an interdisciplinary approach acknowledges that environmental analysis involves a multitude of interconnected disciplines that can only be understood if specialists from these diverse fields interface and work together on common issues. In other words, an interdisciplinary approach requires teamwork. The proliferation of cell phones, the Internet, teleconferencing, and other easy and inexpensive communications technology has enhanced the opportunity for teamwork among specialists, especially those working on generally low-budget EAs who have few resources for travel.

Although nearly all EISs are written by teams of preparers, many EAs for small and simple actions are written by single authors (preparers). Such an approach is fine as long as the preparer is either an expert in the principal issue or issues of concern to the action or consults and receives input from the appropriate specialists. Those specialists may be coworkers on the agency's staff, staff of other agencies, experts with public interest groups, consultants, or subcontractors. Specialists representing those resource area(s) most critical to the analysis should either directly or indirectly participate in, or should be consulted in preparation of the EA.

2.9 Writing documents in plain English

The regulations require agencies to "employ writers of clear prose" preparing analyses that are written in "plain language" so that they can be clearly understood by decision makers and the public (§1502.8).

In the words of one court, a NEPA document is to be "...organized and written so as to be readily understandable by governmental decision makers and by interested non-professional laypersons likely to be affected by actions ..."[9] NEPA has been ahead of the curve with respect to promoting plain English. Decades after enactment of NEPA, the federal government has been promoting plain English throughout reports and other documents to promote transparency. The most recent was an executive order issued by President Obama in 2010.

2.10 Incorporation by reference

The regulations encourage use of incorporating information by reference as a means of reducing the size of an EIS. For example, agencies are required to incorporate existing material into an EIS by reference if it reduces the length of the statement without impeding either the agency's or public's ability to review the document (§1506.3). This is also an appropriate method for reducing the length of EAs.[10] In fact, seeking opportunities for incorporation by reference is even more important for EAs, which ideally should be concise and simple documents, than for EISs.

When material is incorporated by reference, the assessment must reference this material and provide a brief description of its content. All referenced material must be reasonably available for inspection by interested people. Material not publicly available can be added as an appendix to the EA. The availability of the Web has greatly simplified the ability to provide easy access to referenced documents

2.11 Adopting another agency's EA

In an effort to promote efficiency, CEQ encourages agencies to adopt, where appropriate, EISs prepared by other federal agencies (§1500.4[n], §1500.5[h], and §1506.3). However, the regulations are silent concerning the question of whether another agency's EA can be adopted.[11]

The Department of Energy (DOE) has provided the following guidance with respect to this question:

> Any federal agency may adopt any other federal or state agency's EA and is encouraged to do so when such adoption would save time and money. In deciding that adoption is the appropriate course of action, DOE ... must conclude that the EA adequately describes DOE's proposed action and in all other respects is satisfactory for DOE's purposes. ...[12]

Once DOE determines that the originating agency's document is adequate (possibly after adding additional information), DOE is responsible for transmitting the EA to the state(s), Indian tribes, and, as appropriate, the public for preapproval review and comment (unless the originating agency has already conducted an equivalent public involvement process). After considering all comments and complying with the previously mentioned requirements, DOE may issue its own FONSI.

Because the adopting agency is responsible for verifying the adequacy of the analysis and conclusions, it must perform an independent review of the document to be adopted. The EA checklists in Appendix C provide a useful tool for assisting practitioners in performing this review. Because EAs are generally shorter and more narrowly focused than EISs, agencies should pay careful attention to the adequacy of information relevant to their specific role in the action whenever adopting another agency's EA.

2.12 Methodology

As for an EIS, an EA must be accurate, of high quality, and scientifically credible. Completion of the preliminary analyses and internal interim versions of an EA should be followed by a rigorous interdisciplinary peer review by technically qualified reviewers. Comments (internal or external) should be maintained as part of the agency's administrative record.

Chapter two: General concepts and requirements 37

Table 2.4 Performing a rigorous, accurate, and scientific analysis

---

The information must be of *high quality*. *Accurate scientific analysis* ... [is] essential to implementing NEPA.
(§1500.1[b], emphasis added)

Agencies shall insure the professional integrity, including *scientific integrity*. ...They shall identify any methodologies used ...
(§1502.24, emphasis added)

... the analysis is supported by credible *scientific evidence* ...
(§1502.22 [b][4], emphasis added)

... supported by evidence that the agency has made the necessary environmental *analyses*.
(§1502.1, emphasis added)

Environmental impact statements shall be *analytic* rather than encyclopedic. ...
(§1502.2[a], emphasis added)

Emphasis is placed on performing a rigorous, accurate, and scientific analysis that thoroughly investigates potential environmental issues and impacts. Table 2.4 summarizes important regulatory provisions for performing a rigorous, accurate, and scientifically defensible analysis.

2.13 *Fair and objective analysis*

An EA must provide the public and decision maker with a fair, objective, and impartial analysis. Practitioners should strive to avoid even the slightest perception that the analysis may be less than impartial.

In preparing an EIS, agencies must make a concerted effort to "... disclose and discuss ... all major points of view ..." (§1502.9[a]). This requirement is interpreted to also apply to the preparation of EAs. It is therefore recommended that an EA address any reasonable opposing views that have been publicly voiced.

2.14 *Dealing with incomplete and unavailable information*

Special procedures have been established for responding to circumstances involving incomplete or unavailable information when preparing NEPA documents. The challenges of dealing with lack of information can be even more challenging for low-budget EAs than for EISs. Eccleston's companion text, *NEPA and Environmental Planning*, provides a detailed discussion on the requirements for dealing with circumstances involving incomplete and unavailable information.[13]

2.14.1 Incomplete information

If the "incomplete information" is necessary for making an informed choice between alternatives, and the overall cost is not exorbitant, the information must be obtained and included in the analysis (§1502.22[a]). Most agencies' interpretation of "exorbitant" will likely become even tighter as budgets become leaner, especially for generally simple efforts such as preparing EAs. Justification of expenditures for seeking additional data can be even more difficult for simple, low visibility documents such as EAs. Nevertheless, whatever information is needed to support a defensible FONSI should be obtained; the costs are almost always much less than having to proceed to an EIS or defend a FONSI against a lawsuit.

2.14.2 Unavailable information

The agency must clearly indicate when information, relevant to "reasonably foreseeable significant adverse impacts," cannot be obtained either because (1) the overall costs of obtaining it are exorbitant, or (2) the means of obtaining it are not known (§1502.22[b]).

As indicated in Table 2.5, four principal requirements must be satisfied when faced with circumstances involving unavailable information. Where the agency has indicated that information is either incomplete or unavailable, it should at a minimum prepare a qualitative description of the most relevant impacts. However, the reader is cautioned that the inability to satisfactorily quantify an important impact may well render the EA ineffective in supporting a FONSI.[14]

It is recommended that the "rule of reason," in conjunction with a sliding-scale approach, be applied in determining the degree of effort most appropriate for addressing incomplete or unavailable information. Thus the cost and level of effort expended should be commensurate with

Table 2.5 Requirements for dealing with incomplete or unavailable information

1. A statement that such information is incomplete or unavailable;
2. A statement of the relevance of the incomplete or unavailable information to evaluating reasonably foreseeable significant adverse impacts on the human environment;
3. A summary of existing credible scientific evidence which is relevant to evaluating the reasonably foreseeable significant adverse impacts on the human environment; and
4. The agency's evaluation of such impacts based upon theoretical approaches or research methods generally accepted in the scientific community.

(§1502.22[a])

the potential for significance and the value that this information would contribute to decision making.

References

1. Eccleston, C.H., *NEPA and Environmental Planning: Tools, Techniques, and Approaches for Practitioners*, CRC Press, Boca Raton, FL 2008.
2. DOE, *NEPA Lessons Learned*, Issue No. 18, March 1, 1999.
3. Eccleston, C.H., *NEPA and Environmental Planning: Tools, Techniques, and Approaches for Practitioners*, CRC Press, Boca Raton, FL 2008.
4. Eccleston, C.H., *NEPA and Environmental Planning: Tools, Techniques, and Approaches for Practitioners*, CRC Press, Boca Raton, FL 2008.
5. Section 102[2][A] of NEPA, 42 U.S.C. §4332.
6. Blaug, E.A., Use of the Environmental Assessment by Federal Agencies in NEPA Implementation, *The Environmental Professional*, Volume 15, pp. 57–65, 1993.
7. Eccleston, C.H., *NEPA and Environmental Planning: Tools, Techniques, and Approaches for Practitioners*, CRC Press, 2008.
8. Government Institutes Inc., *Environmental Law Handbook*, Chapter 10, 10th edition, 1989.
9. *Oregon Environmental Council v. Kunzman*, 636 F. Supp 632, (U.S.D.C for Oregon 1986).
10. CEQ, *Council on Environmental Quality—Forty Most Asked Questions Concerning CEQ's National Environmental Policy Act Regulations (40 CFR 1500–1508)*, Federal Register, Vol. 46, No. 55, 18026–18038, March 23, 1981, Question number 36a.
11. U.S. Department of Energy, *NEPA Lessons Learned*, pp. 14–15, Issue No. 23, June 1, 2000.
12. U.S. Department of Energy, *Frequently Asked Questions on the Department of Energy's National Environmental Policy Act Regulations*, revised August 1998, Question #15.
13. Eccleston, C.H., *NEPA and Environmental Planning: Tools, Techniques, and Approaches for Practitioners*, CRC Press, Boca Raton, FL. 2008.
14. DOE, *NEPA Lessons Learned*, Issue No. 18, p. 6, March 1, 1999.

chapter three

NEPA and environmental impact analysis

> Predictions are notoriously difficult to make—especially when they concern the future.
>
> —**Mark Twain**

3.1 Introduction

Despite intense effort, the prediction of environmental impacts remains a challenging and inexact art and science. Most National Environmental Policy Act (NEPA) practitioners, who prepare or are otherwise involved with environmental assessments (EAs) as well as environmental impact statements (EISs), have educational backgrounds in engineering or the physical or social sciences. Outside of NEPA, their careers have dealt most heavily with applying scientific facts, theories, models, and equations to solve situational problems. Answers are usually right or wrong; exactness and precision are desired. Impact assessment, whether in an environmental setting (EA or EIS) or otherwise, is, however, a more inexact science than many of these professionals are accustomed to. Impact assessment requires a combination of analytical skills (which are possessed by most scientific professionals) and creative skills (which are not always possessed by graduates of scientific and engineering programs).

Scientists and engineers are usually comfortable with the use of quantitative extrapolation and statistical analysis to perform impact assessment; but they do not always have the insight to discover the multitudes of pathways by which environmental resources can be affected. They can readily apply quantitative models, but first they need to conceptualize each of the pathways requiring application of the models. Overlooking potential impacts can be a problem in EISs, but is particularly easy in EAs, which tend to be shorter and more focused. Whereas many EISs analyze in detail potential impacts to more than ten environmental resource categories, shorter EAs may provide substantial discussion of only one or two resource categories because their proposed action clearly has no potential to affect other resources. While good EAs should not provide a detailed investigation of effects that obviously have

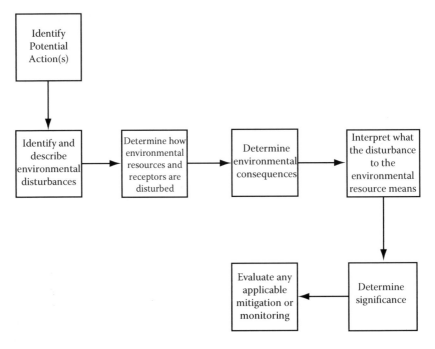

Figure 3.1 The Action-Impact Model. A general-purpose approach for analyzing environmental impacts.

little potential to significantly impact the environment, preparers of EAs must still consider such effects.

The paragraphs below describe a systematic, general-purpose approach for analyzing environmental impacts. The approach is discussed mostly in the context of EA preparation but is also applicable to EIS preparation. This Action-Impact Model is depicted in Figure 3.1. The reader is referred to Chapter 2 in Charles Eccleston's text *Environmental Impact Statements* for a more thorough discussion of environmental impact analysis.[1]

3.2 Actions

Potential actions can be viewed as having several stages of development, one of which is a proposal: "Proposal exists at that stage in the development of an action when an agency subject to the Act [NEPA] has a goal and is actively preparing to make a decision on one or more alternative means of accomplishing that goal and the effects can be meaningfully evaluated" (40 CFR § 1508.23). Once an agency has identified a proposal, it can develop one or more actions to accomplish the objectives of that proposal. For purposes of NEPA, agencies typically identify one action that they feel best accomplishes the objectives of the proposal within the limits of practicality, including the availability of resources such as budget and schedule. That action is commonly termed the proposed action. Other approaches that can accomplish the objectives of the proposal within reasonable limits of practicality are commonly termed alternatives or alternate actions (strictly speaking, the proposed action is one of a group of alternative actions).

Evaluation of alternatives is considered by the Council on Environmental Quality (CEQ) to be the "heart of the EIS" but is not the central objective of an EA. An EA principally serves to provide evidence for determining whether a proposed action could potentially result in significant environmental impacts and therefore require an EIS. If the proposed action would not result in a significant environmental impact, it qualifies for a finding of no significant impact (FONSI) that completes the NEPA compliance process.

However, even though alternatives are not central to an EA, the CEQ regulations still direct agencies preparing EAs to consider reasonable alternatives. CEQ recognizes that the process of considering alternatives might lead to identification of easily implemented alternatives with fewer environmental impacts than the proposed action, even if the original proposed action still falls below the threshold of significance needed to qualify for a FONSI. Hence the ability of an agency to justify a FONSI for an action does not exempt the agency from having to at least contemplate the possibility of accomplishing its objectives using a reasonable and practicable alternative approach that might result in even fewer environmental impacts.

Note that one or more of the alternatives considered in an EA could possibly result in significant environmental impacts; the FONSI is generally for the proposed action, not the alternatives. If the agency can demonstrate that the proposed action would not result in significant environmental impacts, it can issue a FONSI for that action, even if a seemingly practicable alternative evaluated as part of the EA might result in significant impacts. If the analysis performed to prepare an EA indicates that the proposed action might result in significant environmental impacts, the agency cannot generally issue a FONSI for that action, even if a reasonable alternative might avoid significant impacts. The agency can, of course, decide to issue a FONSI selecting such an alternative as its revised course of action. Such a change in selection of an action would be in keeping with the central NEPA objective of informed environmental decision making.

The CEQ regulations clearly note that an EA must include brief discussions of the need for the proposal, of alternatives as required by Section 102(2)(E), and of the environmental impacts of the proposed action and alternatives, as well as a listing of agencies and people consulted (40 CFR 1508.9). With respect to a NEPA analysis, a proposal typically consists of a set of discrete component actions, including any connected actions that are fundamentally inherent in any decision to proceed with an action. For example, a proposed action to build an industrial facility might include distinct connected components such as site preparation, construction of an access road and infrastructure, construction of the plant, and operation of the plant.

All actions reasonably related to the proposal must be identified and adequately evaluated in a section of the EA that describes the proposed action (Figure 3.1). Agencies must be careful not to segment closely interrelated activities into separate NEPA analyses, or they risk challenges. This is particularly true when preparing EAs, where segmented actions might individually not result in significant environmental impacts but where the combined impacts exceed significance thresholds.

Careful definition of the proposal is truly essential before any other elements of the NEPA process may proceed. Too often, agencies begin the mechanical process of NEPA compliance, such as scoping and writing, before having a solid understanding of what they desire to accomplish. In one extreme example, an agency had been preparing an EIS to make a potentially significant change to an ongoing action. But in midstream, a decision was made to cease the action altogether. In response, the EIS had to be altered to address how the agency would not perform the action rather than perform the action. The contractor writing the EIS described the situation as "the proposed action is no action."

3.3 Environmental disturbances

Actions produce environmental disturbances (e.g., air emission, effluents, disruption of flora or fauna, or waste products) that must be identified and described in detail sufficient to support a subsequent analysis of their effect on environmental resources (see Figure 3.1). The environmental disturbances must be identified and characterized in sections of the EA describing the proposed action or environmental consequences. Note that environmental "disturbances" are not in and of themselves environmental impacts unless they affect (adversely or beneficially) environmental receptors.

3.4 Receptors and resources

Environmental disturbances change or perturb one or more receptors (i.e., air or water quality, cultural resources, wildlife, habitat, or human health) that are often referred to as environmental resources (see Figure 3.1). The state of the affected environment (receptors/environmental resources) in an EA is described as it exists before being affected by the proposed action. The CEQ regulations clearly specifies such an approach for an EIS but are silent with respect to an EA. In an EA, the affected environment description can be provided in its own chapter (as in an EIS) or incorporated into the description of environmental impacts. The former approach is usually best reserved for longer and more complicated EAs that include separate written analyses for large numbers of environmental resources. The latter approach is best for short, focused EAs.

3.5 Impact analysis

An EA, like an EIS, must determine the impact on the human environment. To this end, the environmental analysis is directed at determining how environmental disturbances would affect receptors/environmental resources. This analysis is conducted on a resource-by-resource basis in the environmental impact section (sometimes termed "environmental consequences") of the EA (see Figure 3.1). Although the impact analysis in an EA will primarily serve to support a FONSI rather than compare alternatives, as it does in an EIS, the general approach to impact analysis is generally the same in both EAs and EISs.

The result is a set of consequences (i.e., environmental effects or impacts). The terms "impacts" and "effects" are synonymous (§1508.8[b]). The CEQ recognizes three distinct types of impacts (§1508.25[c]): direct, indirect, and cumulative.

The term "effects" is defined to include: "... Ecological (such as the effects on natural resources and on the components, structures, and functioning of affected ecosystems), aesthetic, historic, cultural, economic, social, or health, whether direct, indirect, or cumulative" (§1508.8). Note that the nouns "effects" and "impacts" are synonyms, but the verb synonymous with "impacted" is "affected."

3.6 Significance

Environmental consequences can be considered either significant or nonsignificant, according to specific factors presented in the regulations (§1508.27). Significance is even a greater concern in EAs, which serve principally to support a FONSI, than in EISs, which serve principally to inform decision makers and the public about the environmental consequences of a range of reasonable alternatives. EAs describe environmental consequences in a manner that allows the decision maker to reach a decision regarding their significance or nonsignificance (see Figure 3.1). The reader is directed to Section 4.7 of this text for a more detailed description of the concept of significance. Additional information on significance and its interpretation can be found in *NEPA and Environmental Planning.*[2]

Most EAs do not actually state whether environmental consequences are significant. The purpose of an EA is to provide the technical information needed to support a decision to prepare a FONSI or an EIS, not state what the finding is. Instead, the significance determination is reserved for the FONSI. Some agencies include, however, a draft FONSI near the beginning or end of an EA. Documents that bundle a FONSI with an EA are sometimes termed EA/FONSI documents. Agencies including a FONSI in the same document as an EA must be careful not to merely state that the environmental impacts of its action are not significant without providing the technical basis for that conclusion.

3.7 Mitigation and monitoring

The agency can choose to mitigate potentially significant impacts (see Figure 3.1). The CEQ now accepts the suitability of the "mitigated FONSI," an approach where an agency can use an EA and associated FONSI in lieu of an EIS for a proposal resulting in significant impacts, as long as those impacts are mitigated to nonsignificance. A monitoring program is also an integral element of a well-planned environmental process, particularly where there is a chance that the impact projections could be exceeded. The reader is referred to Chapter 4 in the companion text *Environmental Impact Statements* for a more detailed treatment of this subject.[3]

3.7.1 Mitigation

If the EA concludes that one or more significant impacts would result from the proposed action, the agency might elect to implement mitigation measures that might render a potentially significant impact nonsignificant. An EIS is not required if the potentially significant impacts can be mitigated to the point of nonsignificance.

Methods of mitigation can include avoiding or minimizing the impacts of an action, repairing the effects of impacts that do occur, and compensating for impacts by replacing or substituting resources that have been damaged (§1508.20). If a decision is made to mitigate the impacts to the point of nonsignificance, some agencies prepare Mitigation Action Plans (MAPs) as an integral part of the action. Regardless of whether they prepare a formal MAP, agencies are expected to successfully carry out the mitigation outlined in the EA. Of course, mitigation is an important element of the EIS process as well as the EA process. In an EIS, available mitigation opportunities might not reduce all environmental impacts to nonsignificance, but agencies are expected to identify reasonable mitigation opportunities that could help reduce the magnitude of significant impacts.

The importance of mitigation cannot be overstated. Taking great pains to outline the impacts of an action certainly serves a purpose, but only if that information is then used to determine whether effort might be

Chapter three: NEPA and environmental impact analysis

Table 3.1 Mitigation measures

Mitigation methods may include:
   a. avoiding the impact altogether by not taking a certain action or parts of an action;
   b. minimizing impacts by limiting the degree or magnitude of the action and its implementation;
   c. rectifying the impact by repairing, rehabilitating, or restoring the affected environment;
   d. reducing or eliminating the impact over time by preservation and maintenance operations during the life of the action;
   e. compensating for the impact by replacing or providing substitute resources or environments.

(§1508.20)

practicable to prevent, reduce, or offset those impacts. Later generations are not likely to value the yellowing NEPA documents sitting on agency shelves (or electronically in various types of magnetic media), but they may indeed be very appreciative of the mitigation measures that the earlier generation put in place.

3.7.2 Monitoring

Postmonitoring is an important step in ensuring that environmental predictions are not exceeded and that commitments made in the EA and FONSI are not lost in the haste and confusion of project implementation. Postmonitoring is useful in ensuring that:

1. Environmental standards are met.
2. Mitigation measures are adequately implemented.
3. No impacts are encountered that are substantially different from those originally forecast.

A monitoring and enforcement plan (as part of the mitigation action plan) should be adopted and summarized for any mitigation measures that are chosen (§1505.2[c]). Agencies are also responsible for making the results of relevant monitoring available to the public.

It is important to note that specific performance standards against which mitigation measures can be assessed need to be established. Unless those standards are substantially met, the mitigation has not accomplished its intended purpose. Developing ingenious mitigation measures is highly satisfying, but gains are realized only if the mitigation is actually successful.

References

1. Eccleston, C.H., *Environmental Impact Statements: A Comprehensive Guide to Project and Strategic Planning*, John Wiley & Sons Inc., New York, NY 2000; Eccleston, C.H., *The NEPA Planning Process: A Comprehensive Guide with Emphasis on Efficiency*, John Wiley & Sons Inc., New York, NY 1999.
2. Eccleston, C.H., *The NEPA Planning Process: A Comprehensive Guide with Emphasis on Efficiency*, John Wiley & Sons Inc., New York, NY 1999.
3. Eccleston, C.H., *Environmental Impact Statements: A Comprehensive Guide to Project and Strategic Planning*, John Wiley & Sons Inc., 2000; Eccleston, C.H., *The NEPA Planning Process: A Comprehensive Guide with Emphasis on Efficiency*, John Wiley & Sons Inc., New York, NY 1999.

chapter four

Threshold question
Determining whether an EA or an EIS is required

> Every time I reform in one direction, I go overboard in another.
>
> —Mark Twain

4.1 Introduction

Section 102 of the National Environmental Policy Act (NEPA) requires agencies of the federal government to prepare and include an environmental impact statement (EIS) in every recommendation or report on "... *proposals* for *legislation* and other *major federal actions significantly affecting* the quality of the *human environment*. ..."[1] Often referred to as the threshold question of significance, this mandate provides the linchpin for determining whether an environmental assessment (EA) is sufficient or whether an EIS must be prepared for the proposal. Pay careful attention to the italicized words cited above; each plays a key role in how agencies approach NEPA compliance for specific actions—a process sometimes informally referred to as significance assessment.

Assessing significance is often a formidable task of determining whether the action could result in a significant environmental impact, a determination that would necessitate preparation of a much more rigorous EIS. The outcome of this determination can therefore have profound ramifications in terms of cost, schedule, and potential litigation. Typically, agencies will informally evaluate significance and proceed with preparing a written EA only if their initial evaluation suggests that potentially significant environmental impacts are unlikely, i.e., that the action can qualify for a finding of no significant impact (FONSI). Rarely do agencies publish EAs and then subsequently conclude that a FONSI is not possible and that an EIS is necessary. If initial analytical work on an EA indicates a likelihood of significant environmental impacts, agencies usually decide to alter the proposed action to avoid or otherwise mitigate the significant impacts or transform the in-progress EA document into an EIS (and perform the public notification and scoping requisite to an EIS).

The issue of significance has been the subject of substantial litigation. The threshold question can essentially be dissected into nine discrete criteria that must be satisfied before the EIS requirement as a whole is triggered. Each of these threshold criteria is briefly described in the following sections. An in-depth description of the threshold question is beyond the scope of this book. The reader is referred to the author's (Eccleston) companion book, *NEPA and Environmental Planning Process*, for a comprehensive overview of this subject.[2]

4.2 Proposals

Before an agency can consider the potential significance of an action, it must define it. A proposal might exist in actual fact, even though the agency has not officially declared one to exist. As described in Chapter 1, a proposal is considered to exist either officially or unofficially when (§1508.23):

1. A federal agency has a goal.
2. The agency is actively preparing to make a decision on one or more alternative means of accomplishing the goal.
3. The effects can be meaningfully evaluated.

As noted in Chapter 3, agencies must carefully identify a proposed action that encompasses all of the component activities needed to meet the objectives of their proposal. Identification of an action is more than just characterizing it; it also involves delineating it. Agencies cannot purposefully limit the breadth of their proposed actions in order to qualify for a FONSI knowing that additional interrelated activities will be necessary in order to accomplish the objectives of their proposal (a process sometimes termed *segmentation*).

4.3 Legislation

Proposals can also include submittals for congressional legislation. Specifically, a legislative proposal involves ... "a bill or legislative proposal to Congress developed by or with the significant cooperation and support of a federal agency, but does not include requests for appropriations" (§1508.17).

For the purposes of NEPA, a legislative proposal must be developed by or with "significant cooperation" of a federal agency. The test for "significant cooperation" hinges on whether a proposal is developed predominantly by the federal agency.

4.4 Major

The courts have recognized two distinct interpretations for the term "major." A few courts have interpreted "major" to be a separate criterion, independent of the term "significantly." Under this interpretation, "major" has generally been interpreted as an indicator of the size of a potential action.

In contrast, most courts, and also the Council on Environmental Quality (CEQ), have simply interpreted "major" to reinforce, but not to have a meaning independent from, the term "significantly." Thus under this interpretation, the size of a proposal in and of itself has little bearing on whether the action requires preparation of an EIS. It is the level of environmental impact from a federal proposal that determines whether an EIS is required, not nonenvironmental metrics such as area, cost, or employment (although factors such as area of land affected or employment generated could influence the level of environmental impact). This interpretation avoids dilemmas where a federal action could be deemed major in terms of size or other nonenvironmental metrics, yet would not significantly affect the quality of the human environment; it also avoids dilemmas where a federal action could be deemed small in nonenvironmental terms yet still result in significant environmental impacts.

In the context of NEPA, words like "major" and "significantly" refer to environmental impact and do not necessarily reflect monetary cost. They also do not refer directly to the size of environmental disturbances such as land use or water consumption, although such disturbances are of course indirectly related to the NEPA concepts of "major" and "significant" through their influence on consequent environmental impacts such as land use conflicts or water conservation. NEPA practitioners must constantly look beyond traditional metrics of project importance or significance such as cost and schedule, and focus on environmental impacts.

4.5 Federal

Under NEPA, the term "federal" includes all agencies of the federal government; a federal agency does not include, " ... The Congress, the Judiciary, or the President, including the performance of staff functions for the President in his Executive Office" (§1508.12).

In certain instances, the courts have determined that actions undertaken by a nonfederal agency can be "federalized" for the purposes of NEPA. The amount of federal involvement necessary to "federalize" a nonfederal activity, triggering NEPA, can be a particularly difficult issue to assess. As a broad generalization, nonfederal actions that could not take place "but for" some federal action such as issuing a permit, allocating federal resources (such as federally controlled land or water), or granting federal financial assistance (such as grants or loans) may be "federalized" for purposes of determining the applicability of NEPA. "Federalization" of seemingly nonfederal projects for purposes of NEPA is a complex issue that, like other enforcement issues under NEPA, is often resolved through lawsuits and court decisions. A general-purpose tool for determining when NEPA applies to nonfederal actions is provided in the author's (Eccleston) companion book, *NEPA and Environmental Planning*.[3]

Perhaps the defining court case regarding the universal applicability of NEPA to all federal agencies involved the U.S. Atomic Energy Commission (AEC), which was the forerunner to the U.S. Nuclear Regulatory Commission (NRC). One of the AEC's functions was reviewing and approving applications for licensing construction and operation of new nuclear power plants. The first licensing decision to be made by the AEC following implementation of NEPA was an application by Baltimore Gas and Electric Co. (BG&E) to construct two nuclear reactors in Calvert County, Maryland, approximately 70 miles south of Baltimore. The AEC argued that the action was that of a private entity (the utility applicant) and that the agency was therefore not responsible for NEPA compliance. The court ruled that the action of issuing a license was a federal action subject to NEPA. Federal agencies such as the U.S. Army Corps of Engineers and U.S. Environmental Protection Agency that issue permits to nonfederal applicants now routinely prepare EAs and EISs for their permitting actions. Yet, to this day, some quarters of the NRC maintain that they are not subject to NEPA, and only comply because they wish to.

4.6 Actions

For the purposes of NEPA, the concept of an action is pervasive. Actions include both new and continuing activities. For example, the U.S. Department of Energy prepares NEPA documents for new activities and new facilities but also periodically prepares updated NEPA documents to address decisions to continue ongoing activities at its facilities. Actions also include circumstances where the responsible officials fail to act and that failure to act is reviewable by courts or administrative tribunals under the Administrative Procedure Act or other applicable law.

4.7 Significance

When deciding whether to prepare an EIS, the concept of "significance" is perhaps both the most important and elusive of the nine threshold criteria. Determining the potential significance of an impact can be a daunting task. For this reason, Chapter 8 provides the reader with a rigorous and systematic procedure for assisting decision makers and practitioners in reaching a defensible determination regarding potential significance.

As specified in the regulations, both the intensity and the context in which an impact would occur must be considered. For additional information on significance and its interpretation, the reader is directed to Chapter 8 of the companion book, *The NEPA Planning Process*.[4]

4.7.1 Context

Significance is a function of the setting (i.e., context) in which an impact would occur. The term "context" recognizes potentially affected resources, as well as the location and setting in which an environmental impact would occur.

Four distinct contexts are explicitly identified in the Regulations: (1) society as a whole; (2) the affected region; (3) the affected interests; and (4) the locality. The CEQ regulations for implementing NEPA make it clear that, as a general rule, the significance of an action must be considered in a local rather than global perspective. The NEPA regulations (40 CFR 1508.27[a]) state:

> Context. This means that the significance of an action must be analyzed in several contexts such as society as a whole (human, national), the affected region, the affected interests, and the locality. Significance varies with the setting of the proposed action. For instance, in the case of a site-specific action, significance would usually depend upon the effects in the locale rather than in the world as a whole. Both short- and long-term effects are relevant.

One of the few exceptions to this "localization" rule involves the issue of greenhouse gas emissions, which may affect climate on a global scale. This language discourages agencies from indiscriminately dismissing environmental impacts as not significant based on the argument that nothing dramatic would occur "in the big picture of things" or otherwise on a global level. Otherwise, actions with clearly substantial localized environmental impacts such as aesthetic or noise intrusions into adjoining communities might be interpreted as not significant because the vast majority of the world's population would not experience the intrusions.

Table 4.1 presents examples of categories of environmental effects and the corresponding context that might be appropriate for the assessment of significance. Like most other considerations in NEPA, no formulaic approach can be used for considering context; effective evaluation of context relies on the professional judgment of NEPA practitioners.

Table 4.1 Examples of effects and the corresponding context that might be most appropriate for assessing significance

Specific effect	Corresponding context
Soil impacts	Site and adjoining properties
Wetlands impacts	Site and remainder of subwatershed and watershed
Visual impacts	Viewsheds that include the site
Socioeconomic impacts	Political jurisdictions such as countries and municipalities
Noise	Areas where estimated noise levels generated by the action could be audible
Land Use	Planning district or area covered by comprehensive land use plan

4.7.2 Intensity

Intensity is a measure of the degree or severity of an impact. The regulations define ten factors (i.e., significance factors) that are to be used in assessing intensity (§1508.27[b]). These significance factors are indicated in Table 4.2.

An impact cannot necessarily be deemed nonsignificant simply because the action is temporary. Moreover, agencies are cautioned against segmenting or piecemealing an action by "breaking a project down into smaller component parts" that are individually nonsignificant (§1508.27[b][7]). An example of segmentation involves dissecting a power plant project into a separate analysis of the plant, electric transmission line, gas pipeline, support buildings, and effluent cooling towers. Agencies such as the U.S. Army Corps of Engineers that issue permits for complex or multiphased construction projects are now careful to ensure that the environmental impacts are evaluated holistically rather than segmentally. Agencies must be particularly careful when preparing EAs for construction projects to ensure that the actions are not connected to larger actions whose total environmental impacts could be significant.

4.8 Affecting

An action is said to affect the quality of the human environment. The regulations define "affecting" to mean "will or may have an effect on" (§1508.3). An impact analysis must consider the potential for effects, and not be limited to effects that will certainly occur. The concepts and principles of human health and ecological risk assessment for evaluating adverse effects from environmental contamination, especially in the context of the Comprehensive Environmental Response, Compensation, and Liability Act (CERCLA, commonly referred to as "Superfund"), could

Chapter four: Threshold question 55

Table 4.2 Ten significance factors used in assessing the intensity of an environmental impact (40 CFR § 1508.27[B])

1. Impacts that may be both beneficial and adverse. A significant effect may exist even if the federal agency believes that on balance the effect will be beneficial.
2. The degree to which the proposed action affects public health or safety.
3. Unique characteristics of the geographic area such as proximity to historic or cultural resources, park lands, prime farmlands, wetlands, wild and scenic rivers, or ecologically critical areas.
4. The degree to which the effects on the quality of the human environment are likely to be highly controversial.
5. The degree to which the possible effects on the human environment are highly uncertain or involve unique or unknown risks.
6. The degree to which the action may establish a precedent for future actions with significant effects or represents a decision in principle about a future consideration.
7. Whether the action is related to other actions with individually insignificant but cumulatively significant impacts. Significance exists if it is reasonable to anticipate a cumulatively significant impact on the environment. Significance cannot be avoided by terming an action temporary or by breaking it down into small component parts.
8. The degree to which the action may adversely affect districts, sites, highways, structures, or objects listed in or eligible for listing in the National Register of Historic Places or may cause loss or destruction of significant scientific, cultural, or historical resources.
9. The degree to which the action may adversely affect an endangered or threatened species or its habitat that has been determined to be critical under the Endangered Species Act of 1973.
10. Whether the action threatens a violation of federal, state, or local law or requirements imposed for the protection of the environment.

find applicability in the context of significance determination as well. However, these risk assessment concepts have not to date played a key role in the implementation of NEPA.

With respect to NEPA, an action affects the environment if it produces a change in one or more environmental resources. A reasonably close connection must exist between a disturbance and its resulting effect on the environment.

4.9   *Human environment*

To be significant, an action must substantially affect the quality of the "human environment." As some relationship exists between humans and virtually every aspect of the physical and natural environment, the courts have viewed this term broadly. From a practical standpoint, there is little distinction between the terms "environment" and "human environment." Use of the words "human environment" in the statutory language of NEPA reflects the framers' recognition that good management and stewardship of natural and physical resources benefits the people whose quality of life directly or indirectly depends on those resources.

The "environment" can be divided into two distinct categories: (1) natural and physical environs and (2) manmade or built environs (§1502.16[g], §1508.8[b], §1508.14). Note that cost is not a consideration in NEPA, even though "socioeconomics," the discipline addressing social quality of life issues such as employment and availability of services, is clearly an element of the "human environment" and therefore NEPA consideration. Exclusion of pure cost considerations from NEPA does not suggest that the framers of NEPA considered cost to be unimportant or irrelevant. They simply realized that cost would continue to be considered through processes outside of NEPA, as well as being a key element in developing actions that are practicable. Good NEPA practitioners always remain cognizant of cost considerations and other nonenvironmental practicality concerns throughout the NEPA compliance process, especially when identifying and comparing proposed actions and alternatives. Interestingly, many agencies now prepare "economic impact statements" for legislative proposals that conceptually parallel elements of NEPA.

4.10 Categorical exclusions

Not all federal actions require either an EIS or EA. CEQ encourages each agency to propose categories of related actions that are minor and routine and clearly have little or no potential to result in significant environmental impacts. Agencies propose multiple "categorical exclusions" through rulemaking, giving the public an opportunity to comment. Once a categorical exclusion is adopted, the agency may then determine that certain future actions meet the criteria to qualify under the exclusion, and they then proceed with those actions without preparing an EIS or an EA. Most actions qualifying under a categorical exclusion are very small or simple actions that obviously cannot result in significant environmental impacts. Problems, however, can arise when agencies attempt to "force" actions into qualifying for a categorical exclusion so as to bypass the effort needed to prepare an EA.

A draft memorandum for public comment was issued by the CEQ in 2010. This draft guidance addresses how agencies establish, apply, and review categorical exclusions.* This draft guidance is summarized below.

* CEQ, *Memorandum for Heads of Federal Departments and Agencies: Draft Guidance for NEPA Mitigation and Monitoring, February 18, 2010.*

Chapter four: Threshold question 57

4.10.1 Guidance on applying categorical exclusions

There are two key issues federal agencies face when they want to use a categorical exclusion that has been established and made part of the agency's NEPA implementing procedures:

1. Whether to prepare documentation supporting a categorical exclusion determination.
2. Whether external outreach may be useful in making informed determinations about categorically excluded actions.

4.10.2 Documenting a categorical exclusion

Using a categorical exclusion does not absolve federal agencies from complying with the requirements of other laws, regulations, and policies (e.g., the Endangered Species Act or National Historic Preservation Act). Documentation may be necessary to comply with such requirements. When that is the case, all resource analyses and the results of any consultations or coordination should be included or incorporated by reference in the administrative record developed for the proposed action.

Each federal agency should decide if a categorical exclusion determination warrants preparing separate documentation. There are some activities with little risk of significant environmental effects that generate no practical need or benefit for preparing any additional documentation (e.g., routine personnel actions or purchases of supplies). In those cases, the administrative record for establishing the categorical exclusion may be considered sufficient documentation for applying the categorical exclusion to future actions.

In cases when an agency determines that documentation is appropriate, the extent of the documentation should be related to the type of action involved, the potential for extraordinary circumstances, and compliance requirements for other laws, regulations, and policies. In all circumstances, categorical exclusion documentation should be brief, concise, and to the point. The need for lengthy documentation should raise questions about whether applying the categorical exclusion in a particular situation is appropriate.

If a record is prepared, it should cite the categorical exclusion used and show that the agency determined:

1. The action fits within the category of actions described in the categorical exclusion.
2. There are no extraordinary circumstances that would preclude the project or proposed action from qualifying as a categorically excluded action.

In some cases, courts have required documentation to demonstrate that a federal agency has considered the environmental effects associated with extraordinary circumstances.* Documenting the application of a categorical exclusion can demonstrate that the agency decision to use the categorical exclusion is entitled to deference and should not be disturbed.†

4.10.3 Public engagement and disclosure

Most federal agencies currently do not routinely notify the public when they use a categorical exclusion to meet their NEPA responsibilities. CEQ encourages federal agencies in appropriate circumstances to engage the public in some way (e.g., through notification or disclosure) before using the categorical exclusion.

Agencies should also make use of current technologies to provide the public with access to information on how the agency has complied with NEPA. CEQ recommends agencies provide access to the status of NEPA compliance (e.g., completing environmental review by using a categorical exclusion) on agency websites, particularly in those situations where there is a high public interest in a proposed action.

4.11 Long versus short environmental assessments

CEQ originally intended for EAs to be short documents; detailed descriptions and analyses were originally intended to be reserved for EISs. EAs were never intended to be longer than 15 to 30 pages. In practice, however, a new type of document has emerged in the 30-plus years since issuance of the regulations. This document is the EA of 50 pages or more. Such "long" EAs are typically produced for actions that are somewhat controversial but whose environmental impacts still fall below a threshold of significance. Projects addressed by "long" EAs may have been addressed using EISs had they occurred in the early years of NEPA. But agencies found that they could alleviate public concerns over many controversial projects by demonstrating through a lengthier document, and allowing for public involvement, that the potential impacts would not be significant without all of the procedural rigor of an EIS. There is

---

* Council on Environmental Quality, "The NEPA Task Force Report to the Council on Environmental Quality—Modernizing NEPA Implementation," p. 58, September 2003. Available at http://ceq.hss.doe.gov/ntf/report/index.html.
† The agency determination that an action is categorically excluded may itself be challenged under the Administrative Procedures Act. 5 U.S.C. 702 et seq.

Table 4.3 A comparison between "short" EAs and "long" EAs

Typical "short" EA	"Long" EA
5–10 pages	50–100 pages
General appearance of a long memorandum	General appearance of a short EIS
May closely follow a template	Does not, or may only loosely, follow a template
Stapled	Bound
Simple action	More complex action
Routine action	Nonroutine action
Uncontroversial action	Action may be controversial
Often considers in detail only a Proposed Action and No-Action Alternative	May evaluate 3 or more alternatives in detail
Often combines baseline information and impact analysis	Separate affected environment and environmental consequences sections
FONSI included in same document	FONSI a separate document or bound inside front of EA
Single analyst	Multidisciplinary team of analysts
Usually written by agency staff	Often written by contractor
Usually no formal scoping	Agency may elect to hold a formal scoping meeting similar to one for an EIS
Public may comment, but no formal comment period	CEQ encourages a 30-day comment period for unusual or controversial actions covered in an EA; agencies may elect a longer comment period
Usually only one version	Agency may elect to issue separate draft and final versions, especially if numerous comments are received
Demonstrates no significant environmental impacts	Demonstrates no significant environmental impacts
Supports a FONSI	Supports a FONSI

no official term for these "long" EAs—they are simply termed EAs—and like any EA are usually prepared in the expectation of a FONSI. However, due to their length, preparers can usually convert the document readily to an EIS should analyses suggest that a FONSI may not be justifiable. Table 4.3 outlines some comparisons between "long" EAs and the more typical short EA.

References

1. 42 U.S.C. § 4332 (102)(2)(c).
2. Eccleston, C.H., *NEPA and Environmental Planning: Tools, Techniques, and Approaches for Practitioners*, CRC Press, Boca Raton, FL 2008.
3. Eccleston, C.H., *NEPA and Environmental Planning*, pp. 147–51, CRC Press, Boca Raton, FL 2008.
4. Eccleston, C.H., *The NEPA Planning Process: A Comprehensive Guide with Emphasis on Efficiency*, John Wiley & Sons Inc., New York, NY 1999; *Environmental Impact Statements: A Comprehensive Guide to Project and Strategic Planning*, John Wiley & Sons Inc., New York, NY 2000.

chapter five

Environmental assessment process

> Always acknowledge a fault frankly. This will throw those in authority off their guard and give you opportunity to commit more.
>
> —Mark Twain

5.1 Introduction

The first four chapters of this book have emphasized some key general concepts applicable to the environmental assessment (EA). Before proceeding to the highly detailed discussions of the EA analytical and writing process in the subsequent chapters, a few additional general concepts require some consideration.

To justify issuing a finding of no significant impact (FONSI), an EA must provide clear and convincing evidence that the proposal would not result in any significant environmental impacts, or that any significant impacts can be mitigated to the point of nonsignificance. In contrast, an environmental impact statement (EIS) does not assume this evidentiary burden, as it essentially acknowledges the presence of significant impacts. Consequently, if challenged, an EIS is often easier to defend.

Potential adversaries have taken note of this fact and have refocused efforts on challenging EAs, which are considered more vulnerable. While EISs primarily serve as disclosure documents demonstrating that agencies have taken a hard look at potential environmental impacts and alternatives that might reduce those impacts, EAs must clearly demonstrate that a proposal lacks any potentially significant environmental impact. Just a single potentially significant impact, unless effective mitigation is available, renders an EA incapable of supporting a FONSI. Agencies can counter the trend toward increased scrutiny of their EAs by ensuring that a rigorous investigation has been performed and thoroughly documented.

5.2 Planning the environmental assessment

Figure 5.1 depicts a generalized procedure for preparing an EA. This process must be borne in mind in the initial planning for any EA. A

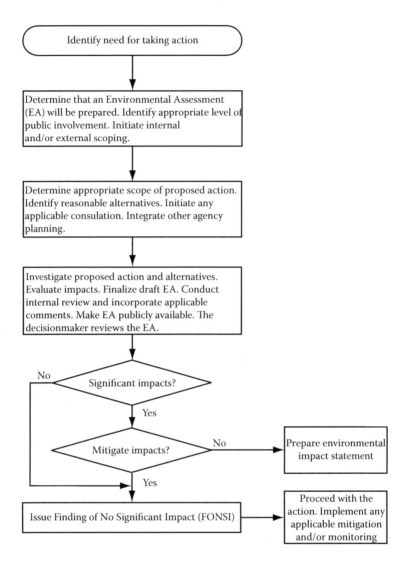

Figure 5.1 Typical environmental assessment process.

National Environmental Policy Act (NEPA) review is typically initiated when a need for taking action has been identified (see oval-shaped box, Figure 5.1). If the action is not eligible for a categorical exclusion (CATEX), the agency's NEPA implementation procedures should be consulted for guidance in determining whether the action is one that normally requires preparation of either an EA or EIS. Many agencies specifically note in their NEPA implementing regulations which categories of actions qualify for a CATEX and which require an EIS, implying that an EA is required for actions not falling into any of the specified categories of actions.

If the agency is uncertain that an action would result in a significant impact, it may choose to prepare an EA to determine whether the action qualifies for a FONSI; this is the course normally taken for many routine and not highly controversial actions, as preparation of an EIS usually requires a substantially larger amount of effort than does an EA. However, as soon as the agency becomes convinced that the action would not likely qualify for a FONSI, it should refocus its efforts on transitioning the EA to an EIS rather than continuing to prepare the EA.

5.2.1 Public involvement

Once a decision to prepare an EA is made, the level of public involvement most appropriate for the proposal must be determined. The stage is now set for initiating internal (and, if applicable, public) scoping (see first rectangle, Figure 5.1).

As warranted, consultations with outside authorities and agencies are also initiated. The importance of this step is witnessed by the fact that one agency reported that the NEPA process was particularly useful in helping the state and tribe resolve their differences regarding the proposed action.[1] The companion text *NEPA and Environmental Planning* provides additional information that may be of use in promoting public involvement.[2] The EA process can be integrated with other planning studies or analyses (see second rectangle, Figure 5.1).

In contrast to an EIS, the CEQ guidelines do not specifically require an agency to incorporate or respond to public comments in an EA. However, as NEPA is intended to be an open public process, at least some effort to seek and respond to public comments is recommended and might be required by some courts. Effort could be made at the scoping stage or after circulation of the EA, or ideally at both times. The ideal expenditure of resources to seek public comment should depend on the complexity and controversy of the action. The agency should decide whether the possible input from the public would help to improve the overall environmental planning process and reduce the likelihood of negative public reaction and lawsuits.

One agency reported that such practice contributed significantly to the success of its EA process. Comments received on the proposed action were placed in the beginning of the EA. References were added to steer the reader to corresponding sections of the EA where the reader could see how these comments were addressed.[3] An effort by the U.S. Forest Service to seek public input during the scoping process for an EA for controlling non-native invasive plants resulted in several comments suggesting that the proposed treatment processes be expanded to additional acreage. The agency evaluated that alternative in the EA and ultimately selected it as its course of action. The result was a program that better met the agency's objectives.

5.2.2   Identifying alternatives

As for an EIS, a systematic process must be performed to identify all reasonable alternatives. Analysts investigate both the proposed action and reasonable alternatives (see third rectangle, Figure 5.1). Alternatives deemed to be unreasonable should be dismissed from further study.

Potentially significant impacts must be rigorously investigated. Once the EA has been finalized, it can then be circulated for internal (and possible public) review; relevant comments are incorporated. The finalized EA can then be made publicly available, and the assessment can be carefully reviewed by the decision maker.

Although the thought process for identifying alternatives for evaluation in an EA is almost always much simpler than for an EIS, the general approach often used for identifying alternatives for EISs can be quickly applied in principle to an EA as well. Of course, using a rigorous and extended process as is common for some EISs is not encouraged for most EAs, but the conceptual principles are still applicable. Picture a "typical" interdisciplinary team assigned to prepare an EA for a relatively large project that does not rise to the threshold for an EIS. While one person is describing the problem, another has already generated the solution, a third is wondering how stakeholders will react, the fourth is questioning how the project will be financed, while the fifth is busy evaluating the environmental impacts. A single preparer rather than an interdisciplinary team may be assigned for an EA for a small or simple action, but that preparer's thought process usually parallels that of a team. It is not uncommon to find that the planning/decision making process lacks a common structured direction.[4]

To address such problems, the U.S. Forest Service has developed a decision protocol process. After obtaining input and developing a pilot process, the Forest Service tested its decision protocol on approximately 20 proposed projects across the country. Decision protocol is based on the observation that a high quality decision is based on the attributes listed in Table 5.1.

In April 1999, the U.S. Forest Service issued its Decision Protocol 2.0.[5] The protocol is a question-based administrative aid that the Forest Service claims can help an interdisciplinary team improve preparation of EAs/EISs. Decision Protocol 2.0 is based on the following five-cycle procedure:

Chapter five: Environmental assessment process

Table 5.1 Attributes of a high quality decision process

The decision process
  Accurately describes the problem and the criteria for solving it.
  Uses available information effectively.
  Collects new information wisely and efficiently.
  Generates and chooses from a wide range of alternatives.
  Distinguishes between facts, myths, values, and unknowns.
  Describes consequences associated with alternative solutions.
  Leads to choices that are consistent with organizational, stakeholder, personal, or other important values.

---

1. Process: The team determines potential decisions that need to be considered, how they can be implemented, and potential constraints (see section later in this chapter titled "Decision-Based Scoping").
2. Problem: The context of the problem is defined through verbal and graphic depictions of the situation, the purpose and need, and definition of the existing information base (including any gaps and uncertainties).
3. Design: This cycle results in a description of the proposal. The team identifies activities to accomplish the objectives, defines alternatives, and identifies cause-and-effect relationships and potential mitigation measures.
4. Consequences: The set of alternatives is refined. Potential impacts are defined. The team considers interactions among other proposed activities, uncertainties, and how design changes could affect key consequences.
5. Action: The team compares alternatives in terms of meeting predefined objectives, cost and reasonability, and avoiding adverse effects. A preferred design is chosen and a defensible rationale is developed and documented. The team examines the sensitivity of this choice in terms of changes in the assumptions. An implementation plan is prepared. As necessary, a mitigation/monitoring plan is also developed.

5.2.3 Determining the potential for significance

As indicated by the first decision diamond in Figure 5.1, two options exist if an agency cannot demonstrate that the potential environmental impacts are clearly nonsignificant: (1) mitigate impacts to the point of nonsignificance; or (2) prepare an EIS. The scope of any mitigation measures must address the range of significant environmental impacts that could occur.

If based on review of the EA, the decision maker concludes that no significant impacts would result (or they can be mitigated to the point of nonsignificance), the agency issues a FONSI; once the FONSI is issued, the agency is free (with respect to NEPA) to pursue the action. The reader should consult the agency's implementation procedures for any variations or additional requirements that must be met.

5.2.4 Three purposes that an EA may serve

Agencies may prepare an EA on any action at any time to further agency planning and decision making (§1501.3[b], §1508.8[a][2], and §1508.8[a][3]).

As depicted in Table 5.2, an EA can serve any one of the three objectives (§1508.9). An EA's most important function is as a screening device for determining whether an action would result in a significant impact requiring preparation of an EIS. Note that in practice the actual screening process usually occurs informally in the initial planning stages, and formal production of a written EA document usually proceeds only after the preparing agency is confident that an EIS will not be necessary. In this way, the first two objectives are closely interrelated. Only rarely will a complete or even a partially complete EA document be used to facilitate preparation of an EIS, following the third objective.

5.2.5 Timing

Preparation of an EA, as for an EIS, must be timed so that it can be circulated with other planning documents (§1501.2[b]).

According to the CEQ, the EA process should normally require a maximum of three months to complete; typically, the process should be completed in less than three months.[6] Such guidance may well be unrealistic, as many agencies report that their EAs typically require more than three months to prepare and issue. Of course, the progress in preparing an EA (or an EIS) is largely dependent on the rate of progress in design and engineering of the project. Delays in the overall design process can translate into corresponding delays in the EA process. Much if not all of these delays is generally out of the control of NEPA practitioners.

Table 5.2 Three purposes of an EA

1. Briefly provide sufficient evidence and analysis for determining whether to prepare an EIS or a FONSI.
2. Aid an agency's compliance with the act when no EIS is necessary.
3. Facilitate preparation of an EIS when one is necessary.

5.2.6 EAs are public documents

Some agencies have operated under the mistaken belief that preparation of an EA does not require public involvement. An EA is in fact a "public document" (§1508.9). Moreover, as defined in the regulations, an EA and FONSI are both included under the definition of an "environmental document" (§1508.10). As an "environmental document," agencies are required to make provisions for public involvement (§1506.6[b]).[7] Agencies are also required to solicit appropriate information from the public (§1506.6[d]).

While EAs must be made publicly available, no formal public-comment review and incorporation period is required by the regulations. Agencies are, therefore, not specifically required to respond to public comments as they are in EISs. While an agency is not required to publicly respond to such comments, it is recommended that the agency at least have some manner of response recorded in its administrative record.

Successfully managing public involvement is one of the keys to effective NEPA practice. A study performed by Reinke and Robitaille sheds light on which public involvement problems have led the courts to find that an EA is inadequate.[8] Problems with EAs typically occurred because the documents had not been made available to the public or had unknowingly excluded interested parties. It is the burden of the preparer to seek out interested or impacted people.

If an agency is preparing an EA to determine whether an EIS is required, other environmental agencies, applicants, and the public must be included to the maximum practical extent in the process (§1501.4[b]).

A scoping process should follow the decision to prepare the EA; this process may take many forms, including internal or public scoping meetings. The following sections discuss the agency's responsibility to include the public in the EA process.

5.2.6.1 Public notification

Agencies are required to make "... diligent efforts to involve the public in preparing and implementing their NEPA procedures (§1506.6[b])." Moreover, agencies are required to provide "public notice" of "... the availability of environmental documents so as to inform those persons and agencies who may be interested or affected (§1506.6[b])." Table 5.3 provides direction on how to notify the public of the availability of either an EA or FONSI (§1506.6[b]).

Although EAs and FONSIs are public documents, they are not required to be filed with the EPA or stored in a national repository, as are EISs.

Table 5.3 Method for notifying the public of the availability of an EA or FONSI

All cases

In all cases, the agency shall mail notice to those who have requested it on an individual action (§1506.6[b][1]).

Where an action is of national concern

In the case of an action with effects of national concern, notice shall include publication in the *Federal Register* and notice by mail to national organizations reasonably expected to be interested in the matter, and may include listing in the 102 Monitor. An agency engaged in rulemaking may provide notice by mail to national organizations that have requested that notice regularly be provided. Agencies shall maintain a list of such organizations (§1506.6[b][2]).

Where an action is of local concern

In the case of an action with effects primarily of local concern, the notice may include (§1506.6[b][3]):

(i) Notice to state and areawide clearinghouses pursuant to OMB Circular A-95 (Revised).
(ii) Notice to Indian tribes when effects may occur on tribal lands or affect cultural resources associated with the tribes.
(iii) Following the affected state's public notice procedures for comparable actions.
(iv) Publication in local newspapers (in papers of general circulation rather than legal papers).
(v) Notice through other local media.
(vi) Notice to potentially interested community organizations including small business associations.
(vii) Publication in newsletters that may be expected to reach potentially interested people.
(viii) Direct mailing to owners and occupants of nearby or affected property.
(ix) Posting of notice onsite and offsite in the area where the action is to be located.

5.2.6.2 Consultation

Federal agencies are required to consult with other agencies, as appropriate, in preparing their NEPA documents. Consultation facilitates a more thorough analysis (§1501.1, §1501.2[d], §1502.25).

Regulatory direction provided in §1502.25 is directed at preparation of EIS but is interpreted to be equally applicable to EAs as well; this is evidenced by the fact that an EA is required to list "... agencies and persons consulted" (1508.9[b]). As a planning vehicle, such a step is particularly useful because it can identify related permitting requirements early in the federal planning cycle. The fact that an action qualifies for a FONSI, or even a CATEX, does not necessarily exempt it from formal consultation

Chapter five: Environmental assessment process

Table 5.4 Public hearings and meetings

A public hearing or meeting is to be held for circumstances involving:
(1) Substantial environmental controversy concerning the proposed action or substantial interest in holding the hearing.
(2) A request for a hearing by another agency with jurisdiction over the action supported by reasons why a hearing will be helpful.
---

requirements under federal statutes such as the Endangered Species Act, National Historic Preservation Act, and Fish and Wildlife Coordination Act.

5.2.6.3 Scoping and public meetings

The CEQ strongly encourages use of a public scoping process to assist agencies in determining the scope of an EA. As indicated in Table 5.4, under some circumstances, a public hearing or meeting may be necessary (§1506.6[c]).

Originally, when the CEQ guidelines were first issued, most actions meeting one or both of the criteria in Table 5.4 were the subject of an EIS, not EA. As more "long" EAs have been prepared for increasingly large and controversial projects, the need for public meetings connected with EAs has increased. The fact that an action lacks potentially significant environmental impacts, and hence qualifies for a FONSI, does not necessarily mean that it is not controversial.

Note that in its NEPA sense, a public meeting is where interested parties, including members of the public, are invited to present comments and questions and express concerns about a proposed action. In some agencies such as the U.S. Nuclear Regulatory Commission (NRC) reserve the term "hearing" to its legal sense as a proceeding presided over by one or more judges where formal testimony is offered. One chief difference between the NEPA public meetings and hearings is that people offering testimony at the hearings must usually meet specified basis and standing thresholds regarding their relationship to the project, while any member of the public is welcome to speak at a NEPA meeting.

5.2.7 *Applicants and environmental assessment contractors*

If an applicant is applying for a federal permit, license, or approval, the responsible agency should begin to prepare the EA "no later than immediately after the application is received" by the agency (§1502.5[b]).

EA contractors: If an agency obtains the services of a contractor to prepare an EIS, the contractor must sign a disclosure statement indicating that it has "no financial or other interest in the outcome of the project" (§1506.5[b]). This regulatory requirement does not apply to a contractor hired to prepare an EA; however, if a contractor is used in preparing an EA, the agency is still responsible for evaluating the environmental issues and must take full responsibility for its scope and content. A safe and prudent approach is for agencies to voluntarily apply the same conflict of interest standards to selection of EA contractors as for EIS contractors.

5.3 Analysis in an EA

The CEQ regulations focus on defining the requirements for performing an EIS analysis. Only limited direction is provided concerning these requirements when an EA is prepared.

5.3.1 Decision-based scoping

To prevent potential disconnects and facilitate more comprehensive planning, analysts should carefully consider the scope of future decision making before initiating the analysis. Specifically, an effort should be mounted to address the following question: "What types of decisions might need to be considered by the decision maker?" The response to this seemingly simple question is essential in shaping the scope and bounds of the analysis.

Chapter 8 of the companion text *NEPA and Environmental Planning* provides a methodology referred to as Decision-Based Scoping (DBS), which is designed to assist practitioners addressing this problem.[9] Under DBS, emphasis is placed on first identifying potential decisions that eventually may need to be considered. Once potential decisions are identified, the scope of actions and alternatives naturally follows. A DBS approach underscores the fact that an EA can be as useful in making decisions regarding implementation of a proposal as it can in its more traditional role of assessing the potentially significant impacts. Note that the full rigor of the DBS approach is rarely necessary or even appropriate for most EAs; however, the conceptual principles underlying DBS are useful considerations for even simple EAs.

To facilitate the DBS task, a tool, referred to as a Decision-Identification Tree (DIT), is also introduced in Chapter 8 of the text *NEPA and Environmental Planning* to assist practitioners in identifying potential decision points. A DIT can be especially useful in scoping complex proposals that may involve numerous engineering, technical, socioeconomic, or regulatory decisions, or where the potential decision points are not clearly obvious. A formal written DIT is rarely necessary or appropriate for most EAs; however, the thought process underlying the DIT can be borne in mind when making decisions in the context of even simple EAs.

5.3.2 Analysis of impacts

As appropriate, potential impacts should be quantified whenever feasible. While the regulations do not specifically state that an EA must consider impacts of connected, similar, and cumulative actions, a determination of nonsignificance cannot be reached without considering such impacts (§1508.27[b][7]).

A FONSI that fails to conclusively demonstrate nonsignificance is highly vulnerable to a successful legal challenge. To this end, the analysis must be rigorous and presented so that a decision maker can clearly reach a conclusion regarding the potential for significance. No stone should be left unturned. The EA should provide specific evidence addressing the significance factors indicated in §1508.27.

5.3.3 Attention centered on proposed action

An EIS must devote "substantial consideration" to each of the analyzed alternatives (§1502.4[b]). In contrast, the EA analysis is normally centered on the proposed action.

Typically, most reasonable alternatives are only briefly described and then dismissed from more detailed study. Sometimes more detailed descriptions of key alternatives and their environmental impacts are appropriate in longer EAs prepared for some of the larger and more controversial actions not requiring an EIS. Discussion is limited to topics that are necessary to give the reader an adequate understanding of what the alternative covers and to allow the reader to discriminate between the alternatives.

Care should be exercised in clearly explaining to the reader why each alternative has been dismissed. Alternatives that have been considered but either not analyzed or rejected are often placed in a separate section labeled "Alternatives Considered but Not Carried Forward."

Such practice can be justified, as the purpose of an EA is generally focused on determining whether an action could result in significant impacts. As long as the proposed action would not significantly affect the environment, there is correspondingly less justification for exploring other alternatives in detail.

While the primary function of EAs is determining the potential significance of a proposed action, it is important to remember that an assessment can also be used to plan actions, consider alternatives, and determine an optimum course of action. In a recent case, for example, a decision had already been made to take action to avoid an environmental fine; an EA was prepared that investigated "all reasonable alternatives." The agency reported that the EA provided an effective planning tool for determining specifically what action to take.[10]

Describing alternatives: A sliding-scale approach may be useful in determining the range of alternatives and the degree to which such alternatives are described; thus projects that are more complex or controversial may necessitate a more thorough investigation. CEQ has recently clarified that a simple range of alternatives constituting a proposed action and no action is often appropriate for many EAs, and that the need for other alternatives is truly based only on the need for practical consideration of competing demands for limited environmental resources.

5.3.4 *Significance determinations are reserved for the FONSI*

While there is not complete agreement, some agencies purport that an EA should be limited to providing decision makers with objective information that they can use in reaching a conclusion regarding significance. Thus the actual determination of significance is made based upon the decision maker's review of the EA. Analysts are therefore confronted with the difficult challenge of preparing an EA that contains information that will clearly lead the reader to a conclusion, yet does not prejudge the potential significance of the impacts.

Accordingly, this text takes the position that the analysis should avoid even the appearance of partiality or the perception of predetermining significance. However, potential impacts that are clearly and unequivocally nonsignificant may be dismissed during the scoping process from more detailed review. If, based on review of the EA, the decision maker reaches a determination that all impacts are nonsignificant (or can be mitigated to such a point), this determination is documented in the FONSI.

Consistent with this approach, judgmental terms such as "nonsignificant" or "acceptable" are best avoided when describing impacts. Less-judgmental terms such as "consequential," "inconsequential," "large," or "small" are more preferable. Better yet, the description of an impact should be quantified whenever feasible.

5.4 Streamlining the EA compliance process

The CEQ encourages use of EAs as a means of streamlining the NEPA process (§1500.4[p] and §1500.5[k]). Although the EA was introduced as a process for streamlining NEPA, the EA process itself may sometimes be improved through streamlining. Some agencies have experienced exceptionally long review and approval cycles, resulting in project delays. The EA review and approval cycle should be examined periodically for inefficiencies. Where practical, reviews involving more than one entity should be conducted in parallel. Agencies should consider delegating approval to the lowest competent decision making level within an organization. Additional methods are described in the following sections for streamlining the EA process.

5.4.1 Reducing the length of an EA

Stating that EAs and EISs are excessively lengthy has become something of a cliché among NEPA practitioners. Reducing EIS length was a key driver in development of the CEQ NEPA regulations in 1978 and, like a movie sequel, returned as a central issue in the proposed NEPA reforms 25 years later in 2003. The very concept of the EA, defined at the outset as a "concise analytical document," is a product of the concern expressed in the early years of NEPA as agencies witnessed the production of lengthy documents for seemingly simple projects. It is therefore ironic that the 2003 redux of the NEPA reforms of the late 1970s placed emphasis on the reduction in the length of EAs! EA length is akin to taxes or the national debt; everyone complains about it but few do anything to change it.

However, not every EA, and indeed not every EIS, is too long. Keeping EAs short is, in general, a laudable objective, but still a secondary objective to supporting a FONSI. Cutting words may be good, but cutting content may not be. The ultimate objective of NEPA is good decisions, not good documents; conciseness serves NEPA's purpose only if it contributes to better communication and hence better decision making. But assuming that the analytical objectives are met, a shorter EA can offer many distinct advantages, including:

- Being easier to read
- Being easier to interpret
- Being less expensive to reproduce
- Taking up less shelf space
- Consuming less paper and other material resources

The first two objectives are paramount. EAs, like EISs, are public documents that must serve to communicate with members of the public. Today's America is a land of images, slogans, and sound bites; the public's attention span is limited. A lengthy EA (or EIS) is like a run-on novel—many people who start reading it will never finish. Most readers lead busy lives and are interested in only certain elements of the analysis. They must be able to rapidly find the material pertinent to their interests. If not, they may overlook the material and instead challenge the FONSI.

The reduced production and storage cost objectives were probably more true in the 1980s and 1990s than today. EAs, like EISs and indeed most government documents, are increasingly published online, where there is no need to consume paper or occupy shelf space. The extra server space needed to house a longer online document costs in the range of pennies. With respect to the preceding paragraph, electronic keyword search capabilities also make it easier for readers of long EAs to find relevant text sections.

Shorter EAs are not necessarily less expensive to prepare than longer ones. Presenting information concisely is an art that takes practice to perfect, and even once perfected it takes work. Many EA writers tend to verbosity followed by sequential rounds of contraction, ultimately culminating under ideal conditions with professional editing. If agency budgets contract in the future, EA preparers may not have the time needed to condense their verbose rough drafts. This fact, combined with reduced pressure to save on reproduction and storage costs, could actually lead to even longer EAs.

Certain actions are available to help achieve conciseness in EAs. First, agencies can perform scoping prior to writing EAs for large or controversial projects. CEQ introduced scoping as a method to narrow the focus of EISs to relevant issues, to discourage the "encyclopedic" character of EISs. All agencies routinely carry out public scoping processes for EISs but not for most EAs. However, there is nothing preventing an agency from holding public scoping for EAs and some do for some of the more controversial actions they cover in EAs. Even if an agency does not wish to perform public scoping, it can still mimic the process internally: visualize how the public might respond if asked about the action and focus the EA on issues estimated to be controversial.

Second, plan the EA carefully. If more than three or four alternatives are contemplated, re-evaluate what truly lies within the range of reasonable alternatives. Although alternatives must receive an analysis technically comparable to that for the proposed action, the text need not be of comparable length. Descriptions of impacts can refer the reader to comparable descriptions presented for the proposed action rather than restating the information. Separate subheadings are not always necessary for each resource analyzed in the environmental consequences text.

Third, consider how to consolidate chapters and sections. Some EA preparers combine the affected environment and environmental consequences chapters; this approach is good as long as both types of information can still be presented clearly. Whether separate or combined chapters are prepared, limit resource-specific sections (subchapters) to those resources requiring detailed analysis to demonstrate nonsignificance. Combine any brief discussion needed for other resources into a single section. Avoid single sentence sections with statements such as "... would not be affected by the proposed action" or "impacts would be as described for. ..."

Fourth, carefully edit the document. Professional technical editors, whose expertise lies with the English language rather than technical environmental fields, can make a substantial contribution. Even if the services of a technical editor are not available, consulting any basic textbook on writing technical prose can help. Strive for short, simple, direct sentences and favor the active over the passive voice. Avoid run-on sentences. Avoid empty clauses such as "it is important that. ..." Try to reduce use of prepositions—you will be amazed at how many words you can eliminate. For example, instead of saying "the stream with high contamination levels" say "the highly contaminated stream." Try to reduce use of that ubiquitous word "the"—it is not always necessary. Eliminating five "thes per page in a 30-page document eliminates 150 words or 450 characters—close to half a page.

The easiest approach to keeping an EA concise is to think in terms of content and coverage, not page length or word count. Think in terms of what topics and what analyses are necessary and what are not. Then describe each retained topic and analysis as concisely as possible without sacrificing content—that and that alone is the appropriate length for the EA: forget about predetermined page lengths.

Incorporating data by reference is a particularly useful method for reducing document length. Focusing on potentially significant issues while reducing attention devoted to nonsignificant issues is another.

A sliding-scale approach may be useful in determining the detail and complexity of analysis that is appropriate. The amount of attention devoted to a given impact increases with the complexity of the proposed action and the potential for significance.

If a decision is made to add material not normally required in an EA, a disclaimer should be included to prevent setting a future precedent. The disclaimer should indicate the reason why inclusion of the material was deemed necessary.

5.4.2 Cooperating with other agencies

To reduce duplication with other environmental requirements, agencies are mandated to cooperate with state and local agencies in preparing their EAs (§1506.2[b][4], §1502.5[b]). Significant savings have been reported when efforts were made to identify and coordinate NEPA with other environmental studies.

5.4.3 Is time money?

In preparing NEPA documents, there was a widely held belief that "time is money." In other words, documents that take an inordinate period to prepare generally cost more than those consuming shorter periods. This premise has recently been challenged. A study of cost versus preparation time has been performed by the U.S. Department of Energy.[11] This study involved 177 EAs prepared between August 1992 and June 1999.

The study found that there is essentially no correlation between document cost and preparation time for EAs. This finding, however, does not suggest that the goal of reducing preparation time is unimportant. Rather, the study suggests that, in striving to reduce cost, one should focus on factors other than preparation time. A related study of EISs reached a similar conclusion.

Instead of concentrating on preparation time for controlling costs, it is recommended that practitioners focus on efficient use of:

- Existing environmental data (e.g., affected environment, designs, accident analyses).
- In-house resources in preparing all or portions of the document.

5.4.4 Tiering

The CEQ strongly advocates a concept referred to as tiering for expediting NEPA compliance. Under tiering, an EA prepared for an action within the scope of an EIS needs only to summarize the issues discussed in the EIS and state where a copy of the statement can be obtained. Discussions presented in the EIS can then be incorporated by reference (§1502.20).

Tiering essentially provides a conceptual or logical basis for viewing agency decision making and implementing actions. Tiering involves the mechanism of "incorporation by reference." However, the mechanism of "incorporation by reference" is not dependent upon "tiering." The

---

TIERING AN EA

"Agencies are encouraged to tier their environmental impact statements to eliminate repetitive discussions of the same issues and to focus on the actual issues ripe for decision at each level of environmental review ... the subsequent ... environmental assessment need only summarize the issues discussed in the broader statement and incorporate discussions from the broader statement by reference and shall concentrate on the issues specific to the subsequent action. The subsequent document shall state where the earlier document is available."

(§1502.20)

regulations are silent on whether an EA can be tiered from an existing EA. The CEQ has informally stated that it is inappropriate to tier one EA from another.[12] However, this is essentially a moot point as an EA can simply incorporate another EA by reference. The ability to incorporate by reference provides essentially the same capability as that provided by tiering.

5.5 Issuing a finding of no significant impact

Four options exist if an agency cannot demonstrate that the impacts of the proposed action are nonsignificant. The agency may either:

1. Mitigate the impacts to the point of nonsignificance.
2. Modify the action to render the impacts nonsignificant (a subset of No. 1 above).
3. Select another alternative (such as no action) that was adequately analyzed in the EA.
4. Prepare an EIS.

5.5.1 FONSI is a public document

Like an EA, a FONSI is a "public document" and must be made publicly available. While the regulations state that the FONSI must be made available to the "affected public" (§1501.4[e][1]), little direction is provided as to how this should be accomplished. The regulations do state that when an action affects a wetland or floodplain special effort should be made to make the FONSI publicly available.[13] For more information, see Section 5.2.6.

5.5.2 Waiting period

Typically, the proposed action may be initiated as soon as the FONSI is issued.

However, under some restricted circumstances (see Table 5.5), a FONSI must be made publicly available for a minimum review period of 30 days before the agency makes a final determination regarding preparation of an EIS and before the action may begin (§1501.4[e][2]). This waiting period is intended to give the public an opportunity to review the FONSI and, if appropriate, take action.

As indicated in Table 5.6, the CEQ also indicates that a 30-day public review period is warranted for circumstances involving:[14]

- Borderline actions.
- Unusual, new types of actions or precedent-setting actions.
- Controversial proposals.

Table 5.5 Circumstances where a 30-day waiting period is required

Under certain circumstances, an agency shall make the FONSI available to the public for a period of 30 days before determining whether an EIS must be prepared and before the action may be taken. These circumstances are:

1. A proposed action is similar to one normally requiring preparation of an EIS under the agency's implementation procedures.

2. The nature of the proposed action is one without precedent.

(§1501.4[e][2])

Table 5.6 Additional CEQ guidance for determining whether a 30-day waiting period is applicable

1. Borderline case, such that there is a reasonable argument for preparation of an EIS.
2. Unusual case, a new kind of action, or a precedent-setting case such as a first intrusion of even minor development into a pristine area.
3. When scientific or public controversy exists over the proposal.

5.6 Administrative record

An agency should not depend solely on environmentally sound decisions as a means of demonstrating compliance with NEPA's requirements. If the agency is challenged, a court may require the agency to turn over files and documentation related to the suit.

Many professionals are typically involved in shaping plans, assumptions, and internal decisions before an EA reaches the decision maker. For the purposes of NEPA, the agency's administrative record can be considered to include the entire record that existed at the time the decision was made, and not simply the portion of the record read by the decision maker.[15] The reader is directed to the companion book *Environmental Impact Statements: A Comprehensive Guide to Project and Strategic Planning* for additional information on developing a thorough administrative record.[16]

5.6.1 Administrative record and case law

Care must be taken to ensure that the determination and resulting FONSI is consistent with the facts and the agency's administrative record. A recent court case illustrates the need for performing a thorough analysis and preparing a comprehensive administrative record. This case involved a state-sponsored action that required federal authorization. Both NEPA and the state's environmental policy act were triggered. The action was later challenged in court. The state was asked to produce the administrative record detailing how the analysis had been performed; it was unable to produce a well-defined administrative record detailing the assumptions used in the analysis and how various decisions had been made. Sensing that it was in a weak position, the state agency agreed to settle the case out of court.

In another case, an EA was prepared for construction of a parking lot in Glacier National Park. The proposed action involved removal of vegetation and 500-year-old cedar trees that the agency's administrative record characterized as "significant in light of the cumulative impacts that have occurred and the extreme rarity of the habitat involved." The EA also contained statements implying that an EIS was needed to consider the project's impacts on unique resources. Further, while other National Park Service analyses identified approximately 200 types of important trees, the original FONSI identified only nine important types of trees that would need to be removed. The court found the FONSI inadequate.[17]

5.7 Serving as an expert witness

Whether you are a client being sued or an environmental practitioner, you may find yourself in the position of serving as an expert witness. Expert witnesses often serve in different capacities. An attorney, for example, may ask an expert witness or an environmental consultant to evaluate the merits of a particular case. An expert in NEPA might also be called upon in a court of law to help prove or disprove a point or the entire case. Expert witnesses serve and support attorneys but do not serve as attorneys themselves. Their job is to be the professionals that they are and bring specialized technical expertise to the courtroom. Expert witnesses will always receive the necessary legal counsel from attorneys assigned to the case.

If you are asked to serve as an expert witness, the first thing you will learn is that there is very little literature written on this subject for the nonlawyer. The following sections are based on an article by M. J. Rogoff that is intended to fill this void.[18]

5.7.1 Are you an expert witness?

An expert witness is generally regarded as a person whose education, skill, or expertise goes beyond the experience of ordinary citizens. If you are asked to serve as an expert witness, you should first check to ensure that there would be no conflict of interest. You should also make sure that you possess the relevant technical expertise specifically required for the case at hand.

The first formal notice of a lawsuit or pleading will typically allege that harm has been done to the plaintiff (the party bringing the suit). The plaintiff bears the burden of proving this allegation. Based on a number of different factors, a judge may deny the plaintiff's claim. An expert can help the lawyer in developing a defensible allegation.

5.7.2 Discovery

Prior to the actual trial, a legal process known as discovery normally allows the plaintiff and defendant to request, obtain, and review information (e.g., business plans, memorandums, e-mails, and technical data) accumulated by the opposing party. Discovery is designed to eliminate surprises by allowing one to mount a defense against charges or evidence that may be entered into trial by the opposing party. A consultant or expert can be a valuable player in helping a lawyer identify and evaluate the types of information that need to be requested from the opposing party.

Experts can be indispensable in helping a judge or jury understand the facts of a case. They can also be particularly effective in sifting through highly technical data and in interpreting the facts of the case.

5.7.3 Depositions and reports

Depositions provide a means of gathering evidence from anyone possessing relevant knowledge about the case. The expert witness is sworn to tell the truth by a court reporter. Expert witnesses might be questioned by the opposing party to identify biases or weaknesses in their testimony. A transcript is prepared documenting the expert witness' statements and responses to questions.

Experts might also be asked to prepare an expert report detailing their findings.

5.7.4 Trial

The expert witness can be indispensable in deciphering the technical mysteries surrounding a case. During cross-examination, the opposing party will attempt to destroy the credibility of the expert. Experts may be confronted with apparent contradictions or errors in their reports of deposition. Experts might even find that their credentials have been brought into question. Table 5.7 provides suggestions to follow in testifying as an expert witness.[18]

When testifying, expert witnesses must exercise care in the way they answer the opposing party's questions. This includes their facial expressions and overall demeanor.

Chapter five: Environmental assessment process 81

Table 5.7 Suggestions to follow when testifying as an expert witness

- Cases and careers have been ended when falsifications were presented in a deposition or at trial. Think before answering a question.
- Take time to think out your answer. At the very least, this will give your attorney time to object to the question or line of reasoning.
- Answer only the question asked. Even if you believe the opposition party's question to be irrelevant, don't follow up with the question you think the examiner should have asked.
- Don't volunteer information. Answer only the question asked, and then stop. You are not responsible for educating the examiner. Once you have answered, remain quiet. The examiner may use this pause and stare at you in a manner designed to intimidate or make you feel uncomfortable; in doing so, the examiner hopes to elicit more information that might be used against your position.
- Don't answer a question you don't understand. It is not your responsibility to ask the question. If you don't understand a question, tell the examiner that you don't understand and ask him to rephrase the question.
- Don't guess. Be as specific as you can, but never guess. If you can't recall, simply tell the examiner that you can't remember.

References

1. U.S. DOE, *NEPA Lessons Learned,* Issue No. 20, p. 17, September 1, 1999.
2. Eccleston, C.H., *Environmental Impact Statements: A Comprehensive Guide to Project and Strategic Planning,* John Wiley & Sons Inc., New York, 2000; and Eccleston, C.H., *The NEPA Planning Process: A Comprehensive Guide with Emphasis on Efficiency,* John Wiley & Sons Inc., New York, 1999.
3. U.S. DOE, *NEPA Lessons Learned,* Issue No. 20, p. 16, September 1, 1999.
4. Berg, J.E., U.S. DOE's *NEPA Lessons Learned,* Issue No. 20, September 1, 1999.
5. U.S. Forest Service, Decision Protocol 2.0, http://www.fs.fed.us/forum/nepa/dp2roadmap.htm.
6. Council on Environmental Quality, *Forty Most Asked Questions Concerning CEQ's National Environmental Policy Act Regulations (40 CFR 1500–1508), Federal Register,* Vol. 46, No. 55, 18026–18038, Question Number 35, March 23, 1981.
7. Blaug, E.A., Use of the Environmental Assessment by Federal Agencies in NEPA Implementation, *The Environmental Professional,* Volume 15, pp. 57–65, 1993.
8. Reinke, D.C. and P. Robitaille, NEPA Litigation 1988–1995: A Detailed Statistical Analysis, *Proceedings of the 22nd Annual Conference of the National Association of Environmental Professionals,* pp. 759–65, 1997.
9. Eccleston, C.H., *Environmental Impact Statements: A Comprehensive Guide to Project and Strategic Planning,* John Wiley & Sons Inc., New York, 2000.
10. DOE, *NEPA Lessons Learned,* Issue No. 18, March 1, 1999.
11. U.S. DOE, *NEPA Lessons Learned,* Issue No. 20, pp. 19–20, September 1, 1999.
12. Personal communications with the CEQ.

13. Executive Order 11988, 42 FR 26951, May 24, 1977; and Executive Order 11990, 42 FR 26961, May 24, 1977.
14. CEQ, Council on Environmental Quality—*Forty Most Asked Questions Concerning CEQ's National Environmental Policy Act Regulations (40 CFR 1500–1508), Federal Register,* Vol. 46, No. 55, 18026–38, March, 23, 1981, Question number 37b.
15. *Haynes v. United States,* 891 F.2d 235 (9th Cir. 1989).
16. Eccleston, C.H., *Environmental Impact Statements: A Comprehensive Guide to Project and Strategic Planning,* John Wiley & Sons, Inc., New York, 2000.
17. *Coalition for Canyon Preservation and Wildlands Center for Preventing Roads v Department of Transportation,* No. CV 98-84-M-DWM, 1999, U.S. District., LEXIS 835 (D. Mont. January 19, 1999).
18. Rogoff, M J., ... So You Want to Be an Expert Witness, *Environmental Practice Journal,* National Association of Environmental Professionals, Vol. 2, Number 1, pp. 5–6, March 2000, ISSN 1466–0466.

chapter six

Environmental impact assessment

> Get your facts first, then you can distort them as you please.
>
> —Mark Twain

6.1 Introduction

This chapter describes the general environmental impact assessment process. Many elements of the process are discussed broadly in this chapter and in greater detail, especially with respect to the EA process, in subsequent chapters.

6.2 Sliding-scale approach

The sliding-scale approach to NEPA analysis can be applied to EAs as well as EISs and may even find greater utility in the realm of EAs. It can make EAs more analytical, briefer, less expensive to prepare, and more effective. It therefore permeates most of the recommendations that follow in this chapter and the next (Chapter 7). This approach recognizes that agency proposals can be characterized as falling somewhere on a continuum of significance with respect to environmental impacts. This approach implements CEQ's instruction that in EISs agencies "focus on significant environmental issues and alternatives" (40 CFR 1502.1) and discuss impacts "in proportion to their significance" (40 CFR 1502.2(b)). Note that under CEQ's regulations and judicial rulings the degree to which environmental effects are likely to be controversial with respect to technical issues is a factor in determining significance. See 40 CFR 1508.27 for guidance on determining significance in the NEPA context.

When applying the sliding-scale approach to performing the analysis for an EA, the preparer should analyze issues and impacts with an amount of detail generally commensurate with their importance. The term "scale" refers to the spectrum of significance of environmental impacts. Proposals with clearly small environmental impacts usually will require less depth and breadth of analysis either in identifying alternatives or analyzing their impacts (though the analysis still must satisfy all

NEPA requirements). Conversely, as proposals fall increasingly closer to the high end of the continuum of potential environmental impacts, the depth and breadth of analysis will increase. Application of the sliding-scale approach is not, however, a rationale for preparing an EA (even a complex EA) rather than an EIS for a proposal with potentially significant environmental impacts. While some EAs need to be more complex than others, proposed actions with the potential for significant environmental impacts normally require an EIS.

6.2.1 Recommendations

- Provide a level of detail and analysis commensurate with the importance of the issue or potential impact. Focus EAs on issues with potential for significant environmental impacts (or that are controversial due to their perceived potential for environmental impacts). Bear in mind that CEQ guidance defines an EA as a concise document that should not contain long descriptions or detailed data (Question 36a, "Forty Most Asked Questions Concerning CEQ's NEPA Regulations," as amended, 51 FR 15618, April 25, 1986; hereafter "CEQ's 40 Questions").
- Identify minor or trivial impacts as such. Include only enough discussion to show why more study is not warranted. If it is obvious to nonexperts that any issue or impact is clearly trivial or irrelevant, the topic need not even be mentioned. If, however, any sector of the general public might object to a topic not being addressed or be confused as to why it is not covered, it is prudent to acknowledge that the topic was considered in the EA process and not overlooked.
- Provide information that a concerned citizen might want, keeping in mind that concerned citizens may need evidence for conclusions that seem obvious to experts preparing the EA.
- Focus analysis on identifying and evaluating potential environmental impacts. Although NEPA requires research, the research should be the minimum necessary to assess environmental impacts. An NEPA document is not a term paper, thesis, journal or review article, textbook, or encyclopedia.
- Remember that affected environment discussions need to contain only the minimum information needed to support the analysis of potential impacts. Following the sliding-scale approach, the depth of affected environment information for an issue (relative to other issues) should be generally proportional to the depth of impact discussion needed for the issue. Remember also that the focus of an NEPA document should be on impacts, not baseline description; aim to keep affected environment descriptions shorter than impact discussions, if possible.

6.3 Identifying alternatives

In general, the range of reasonable alternatives is broader and the number of alternatives appropriately subjected to an analysis of impacts is greater in an EIS than in an EA. (In the following discussion, "analyzed alternative" means an alternative, including any proposed action and no action, for which potential environmental impacts are assessed in detail. It does not include those alternatives considered but dismissed from detailed evaluation.) CEQ's regulations require that an EA include a brief discussion of alternatives to a proposed action as necessary to address unresolved conflicts concerning alternative uses of available resources (40 CFR 1508.9(b)). Although this requirement (which stems from section 102(2)(E) of NEPA) has had varying interpretations, courts are increasingly requiring discussion of reasonable alternatives in EAs. This may reflect the tendency in recent years for many agencies to prepare EAs for actions with impacts of borderline significance that in the past may have been addressed in an EIS.

EAs that address proposals where there is heightened technical controversy surrounding potential impacts or where there is otherwise greater potential for significant environmental impacts from the proposed action may need to evaluate more alternatives than other EAs. Conversely, the smaller the potential for environmental impacts, the less need there is to consider alternatives in detail. In other words, where a proposed action falls on the sliding scale will affect the breadth and depth of the alternatives analysis. All EAs, however, must satisfy minimum requirements as reflected in 40 CFR 1508.9. Alternatives that might have fewer or smaller impacts than the proposed action should be considered. The addition of alternatives to an EA is not a substitute for preparing an EIS when one is required, unless the agency changes its proposal to an alternative lacking potentially significant environmental impacts.

As a general guide for EAs, use the sliding-scale approach when determining how many alternatives to identify and analyze in an EA and the depth of analysis to provide for each alternative. An EA for a very simple proposal might simply limit its range of alternatives to a proposed action and a no-action alternative. Increasingly large or complex proposals will usually require evaluation of more alternatives. Consider adding alternatives if they seem likely to meet the purpose and need of the proposed action, seem practical, and might have fewer or smaller impacts than the proposed action.

As an example, consider this proposal: An agency has a need to treat industrial waste and develops a proposed action consisting of development of a small new waste treatment facility. The site has extensive experience with a particular processing technology and expects to use that technology in the new facility. Other technologies might, however, be applicable, but they have received less research and development and no funding for further development. An EA might consider alternatives that use other technologies, particularly those that might have environmental, safety, or cost advantages. The EA would not, however, need to contain detailed evaluation of impacts from technologies that are clearly infeasible, inappropriate, or unproven.

6.3.1 Types of alternatives

The CEQ directs that EAs consider a no-action alternative; identification of other alternatives is a more subjective process relying on the judgment of the preparers and considerations of reasonableness and practicality. Although it is impossible to categorize all types of alternatives potentially considered in an EA, the following broad categories of alternatives are common to many EAs and EISs:

- Site alternatives—implementing an action, usually a construction project, at differing locations.
- Route alternatives—similar to site alternatives, but construction of linear projects such as transmission lines, pipelines, and highways following different alignments.
- Intensity alternatives—application of differing levels of the same action.
- Technology alternatives—implementing differing technical actions to accomplish a stated objective.
- Schedule alternatives—implementing an action at differing times or components of an action in differing chronological order.

Other categories of alternatives are possible. It is generally unnecessary to encompass all of the above alternative categories in a range of alternatives in an EIS, and especially in an EA. Remember that scoping, whether formal or just informal in the minds of the preparers and agency staff, is available as a process to help define alternatives as well as to narrow impact analyses. The concept of having site and route alternatives is intuitively obvious for construction projects, and the mention of the word "alternative" automatically triggers thoughts of such location-based alternatives among many NEPA practitioners. But such a simplified approach can be inadequate for NEPA, even for an EA. For example, an EA (or EIS) for a power plant may have to consider alternatives based on construction technology or fuel source.

6.3.2 Range of reasonable alternatives

The CEQ NEPA regulations (40 CFR 1502.14) provide guidance for preparing the chapter on the proposed action and alternatives in an EIS. While

40 CFR 1502.14 is focused on preparing an EIS, it generally also applies to preparation of EAs.

1. Include reasonable alternatives not within the jurisdiction of the lead agency.
2. Include the no-action alternative. The no-action alternative is the most likely future that could be expected to occur in the absence of the project. Where the future is different from existing conditions, the differences should be clearly defined.
3. Include appropriate mitigation measures not already included in the proposed action or alternatives.

CEQ's regulations require that an EA include a brief discussion of alternatives to a proposed action that involves unresolved conflicts concerning alternative uses of available resources (40 CFR 1508.9(b)). Although this requirement (which stems from section 102(2)(E) of NEPA) has had varying interpretations, courts are increasingly requiring discussion of alternatives in EAs, especially longer and more complex EAs. In addition, one agency's (Department of Energy [DOE]) NEPA regulations (but not CEQ's regulations) require that EAs include a no-action alternative (10 CFR 1021.321(c)). Thus a DOE EA is used to satisfy agency planning requirements for analyzing alternatives, as well as to determine whether to prepare an EIS.

EAs that address proposals where there is heightened technical controversy surrounding potential impacts or where there is otherwise greater potential for significant environmental impacts from the proposed action may need to identify and analyze more alternatives than other EAs. Conversely, the smaller the impacts of the proposed action, the less need there is to consider alternatives. In other words, where a proposed action falls on the sliding scale will affect the alternatives analysis. All EAs for DOE, however, must satisfy minimum requirements as reflected in 40 CFR 1508.9 and 10 CFR 1021.321. Alternatives that might have fewer or smaller impacts than the proposed action should be considered. The addition of alternatives to an EA is not a substitute for preparing an EIS when one is required.

Identify and briefly discuss alternatives considered but dismissed from detailed evaluation, particularly any such alternatives raised during the public scoping or comment process. EA preparers should explain why an alternative is not reasonable (e.g., that the cost or time to implement would be impractical or technical implementation would be infeasible) and make the method for screening alternatives clear to readers.

6.3.2.1 Appropriate number of alternatives to include in the analysis

The number of alternatives that is considered reasonable generally varies depending on the nature of the purpose and need for the action. The alternatives described in a chapter should be representative of all of those possible actions that can be reasonably expected to satisfy the purpose and need.

In some scenarios, such as fishery management, there may be an infinite number of alternatives to satisfy the purpose and need. This is particularly true when the purpose and need is fairly broad. For example, an agency may consider an entirely open fishery with no controls, close the fishery entirely, or any combination of partial closures.

CEQ's "Forty Most Asked Questions," Question 1b states that for some proposals there may exist a very large or even an infinite number of possible reasonable alternatives. When there are potentially a very large number of alternatives, only a reasonable number, covering the full spectrum of alternatives, must be analyzed and compared in the EIS. This requirement for analyzing all reasonable alternatives also applies for EAs. What constitutes a reasonable range of alternatives depends on the nature of the proposal and the facts in each case.

One may mathematically think of two basic scenarios for possible ranges of alternatives: discrete alternative ranges that comprise two or more qualitatively distinct alternatives, and continuous alternative ranges that comprise a broad spectrum of qualitatively similar but quantitatively differing alternatives. An EA considering only a proposed action and a no-action alternative is a simple example of a discrete alternative range. A range of possible fuel sources for power generation, e.g., coal, natural gas, nuclear, wind, and solar, would be a discrete range of alternatives. A range of acreage for applying a proposed pest management practice on public land would be a continuous range of alternatives. A reasonable approach might be to address alternatives that apply the procedure to the entire land tract, half of the land tract, and none of the land tract (no action). Of course, many EISs and some EAs must consider a combination of alternatives drawn from both discrete and continuous ranges, e.g., two alternative pest management procedures applied over differing extents of a tract of land.

6.3.2.2 Art of defining a range of reasonable alternatives

Defining reasonable alternatives is an art form in NEPA practice whose mastery is critical to achieving effective NEPA analyses. This is especially true when having to draw from continuous ranges of alternatives, but can sometimes be challenging when drawing from discrete ranges. Analyzing only a proposed action and a no-action alternative, as occurs frequently in EAs, is often (but not always) inadequate. Although the CEQ guidelines do not specifically state that additional alternatives are always needed, they do call for consideration of a reasonable range of alternatives. Yet detailed consideration of more than three or four alternatives defeats the conciseness objective for an EA. The challenge can seem daunting: a combination of 3 possible locations, 3 possible technologies, and 3 possible fuel sources for a hypothetical power plant project results in 27 alternatives! The no-action alternative brings that number to 28. This is far too many alternatives even for an EIS, let alone an EA. And this example involves only discrete ranges of alternatives! The numbers for an example involving continuous as well as discrete ranges of alternatives could be even more daunting.

The best approach is to use an informal screening process to rapidly narrow a broad field of possible alternatives to a few worthy of detailed analyses. Two EAs prepared for electric transmission line construction projects in southeastern Virginia illustrate an effective screening process for route alternatives. For each EA, the contractor developed a map of the study area depicting a grid of possible route segments, termed "links," for constructing the transmission line. The contractor then identified more than a dozen linear combinations of links, termed "routes," that allowed for construction of the entire transmission line from the desired start and end points. Each route could then be scored based on its overall proximity to sensitive environmental resources. In each case, the four highest scoring routes were identified as alternatives for detailed analysis in the EA text. The other alternatives were presented graphically in the EA but were dismissed in the text without further verbal analysis. Despite documenting more than a dozen alternative routes, the EA text was only about 50 pages long.

The benefits of alternatives analysis are ideally achieved before the document is even written. A non-native plant control EA for a forest initially considered, in addition to the proposed action and no-action alternative, two additional alternatives. The proposed action was implementing a combination of physical control methods, herbicide applications, and release of certain weevil species known to feed upon and kill specific targeted plant species. One alternative was to increase the areas subject to each type a treatment—an intensity alternative. Another was to exclude use of the weevils. Many comments received during scoping questioned why the agency was not proposing an even more aggressive control program. In response, the agency decided to implement the increased intensity alternative. Although choice of the original proposed action would have supported a finding of no significant impact (FONSI), choice of the alternative both supported a FONSI and better accomplished the stated purpose and need.

6.3.2.3 Comparing alternatives

An objective comparison of the reasonable alternatives is an essential feature of every EA. The environmental consequences discussion should compare the impacts of the alternatives and provide a simple mechanism for the reader to compare the alternatives. An alternatives comparison table is a simple way to show the impacts of all of the alternatives. A table followed by a few sentences of discussion may be adequate for simple EAs; larger or more complex EAs might also require written descriptions of the impacts from each alternative.

Table 6.1 shows a simplified example alternatives comparison table. The table shows the impacts to resources by alternative. The table used in an EA may have more detail and should address all of the resources described and analyzed in the EA.

6.3.2.4 General recommendations

- Do not overlook reasonable technology, transportation, or siting alternatives, including offsite alternatives.
- As a general guide for EAs (for which requirements are less specific than for EISs), use the sliding-scale approach when determining how many alternatives to identify and analyze in an EA and the depth of analysis to provide for each alternative. Consider adding alternatives if they might have fewer or smaller impacts than the proposed action.
- Alternatives should be defined broadly enough (i.e., be "robust" enough) to allow small changes in the way the agency implements the selected alternative, but not so broadly to preclude meaningful

Table 6.1 Example alternatives comparison table

Resource	Alternative 1 no-action	Alternative 2 proposed action	Alternative 3
Aquatic Habitat	No impacts.	No significant impacts, some minor beneficial impacts.	Same as preferred alternative.
Wildlife	No impacts.	Minor beneficial impacts for riparian-dependent wildlife.	No impacts.
Vegetation	No impacts.	Minor beneficial impact of more native vegetation in riparian areas.	Minor impacts, but none at the watershed scale.
Soils	No impacts.	Same as no-action, except practices would focus on prevention of soil movement into salmonid habitat. Practices should eliminate soil structure impacts.	Same as preferred alternative.

analysis. Most NEPA documents, EISs as well as EAs, are prepared based only on preliminary or conceptual designs; the final designs typically involve modifications to many of the fine details. A good NEPA practitioner will describe each alternative in broad enough terms that serve to "envelope" the possible environmental impacts in a way that anticipates a "worst case" maturation of a design. Otherwise the agency might find it necessary to later supplement an EA or EIS; and in the case of an EA, the supplemental analysis might possibly indicate possibly significant environmental impacts thereby necessitating an EIS.
- Identify the range of reasonable alternatives that satisfies the agency's purpose and need. Especially for simple proposals, a suitable range might consist only of a proposed action and no-action alternative; more complex proposals may require more alternatives (i.e., more than one "action" alternative).

6.3.3 No-action alternative

The no-action alternative may or may not be a reasonable alternative. Even if it is not, it serves one particularly valuable purpose. It provides a baseline against which impacts of the other analyzed alternatives can be compared. No action does not necessarily mean doing nothing. Instead, it often involves maintaining or continuing the "status quo" (e.g., a management plan or activity covered by an existing NEPA review). For proposed new projects, no action means that the proposed activity would not take place. For proposed changes to an ongoing activity, no action is best considered to mean continuing with present plans. Some NEPA documents also use no action to mean discontinuing the present course of action by phasing out operations, although strictly speaking such a discontinuance of activity would constitute an action alternative.

6.3.4 Recommendations for identifying and investigating alternatives

Recommendations for identifying and investigating alternatives are detailed in Table 6.2.

6.4 Impacts

As for choice of alternatives, impact analysis should generally be less detailed in EAs than in EISs, and more detailed in EAs for more complex or potentially controversial environmental impacts than in EAs for simple actions. That is, the sliding-scale approach applies to impacts analysis in

Table 6.2 Recommendations for identifying and investigating alternatives

- Investigate a range of reasonable alternatives that satisfies the agency's purpose and need. Include alternatives that would respond to the underlying purpose and need under a variety of reasonably foreseeable circumstances.
- Address *reasonable* alternatives that are outside the agency's jurisdiction, even if they conflict with lawfully established requirements (e.g., an alternative that could be reasonable if an existing law could be amended or if a regulatory agency granted a waiver).
- Briefly describe alternatives considered but dismissed from detailed evaluation, particularly any such alternatives raised during the public scoping or comment process. Failure to consider alternatives that *seem* reasonable to others may affect the credibility of an otherwise adequate NEPA review.
- Explain why an alternative is not reasonable (e.g., that the cost or time to implement would be impractical, technical implementation would be infeasible). Make the method for screening alternatives clear to readers.
- Investigate alternatives that seem impractical only because of current assumptions, but otherwise would be reasonable. Consider whether it is foreseeable that technical or economic factors might change such that an apparently infeasible but otherwise reasonable alternative would become feasible.
- Note that infeasible alternatives are certainly unreasonable; feasible alternatives also may be unreasonable. Example: It might be feasible to build a new facility at a given site without regard to infrastructure because all necessary infrastructure already exists. It might not be a reasonable alternative, however, to build the same facility at another site because the required infrastructure is not in place and could not be provided at reasonable time or expense.
- Do not overlook reasonable technology, transportation, or siting alternatives, including offsite alternatives.
- Consider the no-action alternative even if an agency is mandated a legislative command to act. Include discussion of the legal ramifications of no action, if appropriate.
- In some situations, no action taken by an agency may constitute the only alternative to the proposed action.
- When an NEPA document addresses ongoing activities, it may be useful to consider more than one version of no action. For example, a forestwide EIS might analyze the continuation of the present course of action as one no-action alternative and discontinuation of present operations as a second no-action alternative. In such cases, it is important to carefully name the no-action alternatives to avoid confusion.

much the same way as it applies to the choice of alternatives. An unusually complex and lengthy EA, however, may indicate the need for an EIS (Question 36b, "CEQ's Forty Questions").

An impact may be adverse or beneficial, and the overall impacts of an alternative may be significant even if on balance the impacts would be beneficial (40 CFR 1508.8 and 1508.27(b)(1)). In general, impacts (beneficial and adverse) are more thoroughly analyzed in an EIS than in an EA because an EIS deals with actions that admittedly may have significant impacts. Section 102(2)(C) of NEPA requires evaluation of:

- Unavoidable impacts
- The relationship between short-term uses of the environment and the maintenance and enhancement of long-term productivity
- Any irreversible or irretrievable commitments of resources

In addition, CEQ regulations (40 CFR 1502.16) state that an EIS should discuss:

- Direct and indirect effects
- Possible conflicts between the proposed action and the objectives of federal, regional, state, local, and tribal land use plans, policies, and controls
- Energy and natural or depletable resource requirements and conservation potential of alternatives and mitigation measures
- Urban quality, historic and cultural resources, and the design of the built environment

CEQ regulations also require consideration of cumulative impacts (40 CFR 1508.25(c); cumulative impacts are discussed in Section 6.8.4.4). An impact may be adverse or beneficial, and the overall impacts of an alternative may be significant even if on balance the impacts would be beneficial (40 CFR 1508.8 and 1508.27(b)(1)). In addition to characterizing impacts, an EA or EIS should discuss means to mitigate impacts.

6.4.1 Bounding analysis

A bounding analysis is an analysis designed to identify the range of potential impacts or risks, both upper and lower. Such an approach might be used in an EA to simplify assumptions, address uncertainty, or because expected values are unknown. As a practical matter, a bounding analysis most often is used to provide conservatism in the face of uncertainty. The greater the uncertainty in an impact analysis, the greater is the need for reliance on conservative assumptions. Generally, if the greater level of impacts resulting from the conservative assumptions is not significant, lower levels can be expected to also not be significant.

CEQ distinguishes between the "environmental consequences section" of an EIS, which should be devoted largely to a scientific analysis of the impacts of the analyzed alternatives,* and the "alternatives section," which should present a concise comparison of alternatives. In some cases, it is considered good practice to make such a separate comparison in an EA as well.

6.4.2 Analysis of impacts

In general, impacts (beneficial and adverse) must be more thoroughly analyzed in an EIS than in an EA because an EIS deals with actions that are assumed or admitted to have significant impacts. An EIS must devote substantial treatment to all reasonable alternatives analyzed to enable a comparison among those alternatives. An EA generally must only include sufficient analysis to demonstrate that the environmental effects from a project do not pose potentially significant impacts. All NEPA documents including EAs, regardless of length or complexity, must rely on high quality information, including accurate scientific analysis (40 CFR 1500.1(b)).

An EA may focus the impacts analysis on the proposed action in order to provide the basis for a significance determination. The CEQ regulations (40 CFR 1508.9(b)), however, require an EA also to include brief discussions of the impacts of alternatives to a proposed action that involves unresolved conflicts concerning alternative uses of available resources. The preparer of an EA should carefully consider the potential for impacts from alternatives, but the discussion of impacts from each alternative need not be equally long or detailed as that for the proposed action. Likewise, the discussion for each alternative need not be equally long. A brief indication as to why an alternative is not clearly environmentally preferable to the proposed action may sometimes suffice. As with the choice of alternatives (Section 4.2), the analysis of impacts of alternatives should be more detailed in an EA where there is heightened technical controversy surrounding potential impacts or where there is otherwise greater potential for significant impacts. That is, the sliding-scale approach applies to impacts analysis in much the same way as it applies to the choice of alternatives. An unusually complex and lengthy EA, however, may indicate the need for an EIS.†

When confronted with incomplete or unavailable information for an analysis of reasonably foreseeable, significant adverse environmental effects, CEQ's regulations require agencies to indicate that such information is lacking and to obtain the information when doing so does not entail an exorbitant cost and the information is essential to a reasoned choice

---

* CEQ, "CEQ's Forty Questions," Question 7.
† CEQ, "CEQ's Forty Questions," Question 36b.

among alternatives (40 CFR 1502.22). Otherwise, agencies are to discuss the implications for the analysis. The fact that critical information needed to assess the significance of an impact is not available and can not be obtained through a reasonable effort may be a sufficient basis to conclude that the impact is potentially significant, triggering the need to prepare an EIS.

Recognize that relative comparisons do not provide absolute information, and both relative and absolute information generally are needed. For example, the statement "routine emissions would increase by 0.05 percent" does not describe an impact (although it is a valuable part of the description of the alternative). The statement provides neither the absolute value of emissions nor rigorous basis for determining their environmental impacts. Further, relative comparisons, particularly those given without a baseline of absolute magnitude, may be misleading (e.g., "99.9% pure water" could describe sewage). NEPA documents should always carry a factual, unbiased tone, and never rely on the verbal sleight of hand common to advertising or political sound bites.

Both the estimated frequency and the magnitude of impacts inform readers about the nature of the impact and mitigation options. Often, the most significant impacts arise from scenarios that are unlikely to occur, whereas less significant impacts may occur frequently.

In general, use available data for an EA. Most federal installations with a substantial history of environmentally detrimental activity have accumulated an extensive number of environmental reports, databases, and maps. Today, an abundance of data is available online, often as geographical information system (GIS) layers, even for locations that have not been the subject of previous environmental reporting. If data needed to quantify impacts are not available, qualitatively describe the most relevant impacts. Be aware that inability to satisfactorily characterize an important impact in an EA likely will render it inadequate to support a FONSI.

The following paragraphs discuss general analytical approaches for some of the more common environmental issues covered in EAs (and EISs). They serve an instructive purpose only, providing examples of a general approach to impact analysis in NEPA documents, especially EAs. It is not however a comprehensive encyclopedia of all of the issues and types of environmental impacts commonly encountered in the preparation of EAs and EISs.

6.4.2.1 Biological impacts

Few actions other than purely administrative ones have no possible effect on plant or animal (wildlife) species. Unless within completely paved areas, most actions involving construction of new, expanded, or modified facilities inevitably require permanent loss of natural habitat that provides a home for plant and wildlife species. Most actions involving modification to streams or other bodies of water, or to land contributing surface runoff to those water bodies (sometimes termed the watershed), inevitably affect populations of fish, invertebrates (many are aquatic life stages of insects), plankton (free-floating microscopic organisms), and birds and mammals that feed upon these aquatic species. Some activities may expose plants and animals populations (biota) to radioactive or chemically hazardous materials in environmental media or to materials released in waste streams. Either individually or in any combination, physical, chemical, or radiological stressors can impact biota. A variety of risk and dose assessment tools are available for evaluating potential impacts to biota, and for performing dose and risk assessments.

In most circumstances, the principal biological issues of concern are assessed at the population level, with a focus on the potential decrease in local species population size and a decline in suitable habitat. However, where threatened or endangered species, or commercially or culturally valuable species, may be affected, consideration of impacts to individuals may likewise be of importance (Table 6.3).

6.4.3 Incomplete or unavailable information

When confronted with incomplete or unavailable information for an analysis of reasonably foreseeable, significant adverse environmental effects,

Table 6.3 Recommendations for assessing biological impacts

- Analyze the impact of each alternative on aquatic and terrestrial biota, as appropriate. In addition to chemical and radiological stressors, consider physical stressors individually and cumulatively, when assessing potential biological impacts. For example, a physical stressor might be the physical impact a wind farm could have on migratory birds.
- Consider possible effects on each natural habitat in the affected area.
- Identify impacts on federal and state threatened or endangered species and their critical habitat. Consider also species proposed for federal or state listing, as well as federal candidate species.
- Recognize that routes of exposure for biota and people may differ. For example, human access to a hazardous landfill might be restricted, but the restrictions might not prevent exposure to animal and plant life. Exposure pathways from contaminated soil or sediment for a burrowing terrestrial animal and benthic aquatic animal, respectively, could be much greater than for humans.
- For actions involving possible releases or handling of hazardous materials or petroleum products, consider whether implementation of each alternative would comply with applicable dose limits for protection of biota.
- When necessary to perform biota dose and risk assessments, identify the appropriate dose/risk standards, criteria, or benchmarks. When conducting assessments, use principal indicator species in the analysis.

CEQ's regulations require agencies to indicate that such information is lacking and to obtain the information when doing so does not entail an exorbitant cost and the information is essential to a reasoned choice among alternatives (40 CFR 1502.22).

Where the uncertainty is significant or a major factor in understanding the impacts, explain how the uncertainty affects the analysis. Be sure to identify any responsible opposing views regarding how to conduct impacts analysis or interpret conclusions.

The EA should (1) acknowledge the controversy (i.e., the differences of opinion or fact) and (2) explain the basis for DOE's choice of methodology in a manner that demonstrates that the analysis in the EA is technically sound and provides a sufficient basis for decision making. Also, consider explaining how use of the different methodology would affect the conclusions, as this different perspective might be useful to the decision making process. In some cases, it may be prudent and useful to present the results of using the alternative data, assumptions, or methodologies.

6.4.4 Cumulative effects analysis in EAs

The combined incremental effects of human activity pose a serious threat to environmental quality. While such impacts may be nonsignificant by themselves, over time they can compound, from one or more sources, to the point where they significantly degrade environmental resources. The analysis of cumulative effects provides a powerful tool for taking such effects into account as federal officials plan future actions. Consideration of cumulative impacts is especially controversial in the context of EAs, where cumulatively significant impacts could prevent issuance of a FONSI for actions whose individual impacts might all be insignificant.

In recent years, the issue of cumulative impacts has become the subject of increasing interest and litigation. This section has been prepared to assist practitioners in preparing a cumulative impact analysis that meets analytical and regulatory requirements. For a more in-depth discussion of the analytical and regulatory requirements, the reader is referred to the companion texts, *The NEPA Planning Process* and *Environmental Impact Statements*.[1]

A "cumulative impact" is defined as:

> ... the impact on the environment that results from the incremental impact of the action when added to other past, present, and reasonably foreseeable future actions, regardless of which agency (federal or non-federal) or person undertakes such other actions. Cumulative impacts can result from individually minor but collectively significant actions taking place over a period of time (§1508.7).

6.4.4.1 Purpose of the analysis

By mandating an analysis of cumulative impacts, the regulations ensure that the range of actions considered in NEPA documents takes into account not only the project proposal but also other actions that could cumulatively harm environmental quality. The results of the cumulative impact analysis should be incorporated into the agency's overall environmental planning. The conclusions can also be incorporated into the plans of state and private entities. To this end, federal agencies are to use results obtained from the cumulative impact analysis as a tool for evaluating the implications of a proposal in even project-specific EAs.[2]

Emphasis must be placed on identifying activities occurring outside the jurisdiction of their respective agencies. However, the analysis should be limited to considering those past, present, and future actions that incrementally contribute to the cumulative effects on resources affected by the proposed action; actions affecting resources that have sustained cumulatively insignificant impacts generally do not add to the value of the analysis.[3]

6.4.4.2 Performing the analysis

This section briefly summarizes the process used in performing a cumulative impact analysis.

6.4.4.2.1 Components of an adequate cumulative impact analysis A cumulative effects study must identify:[4]

- The specific area in which effects of the proposed project would be felt.
- Impacts that are expected in that area from the proposal.
- Other past, proposed, and reasonably foreseeable future actions that have impacted or could be expected to impact the same area.
- Expected impacts from these other actions.
- Overall expected impact if the individual impacts were allowed to accumulate.

In one case, a court deferred judgment in favor of the U.S. Forest Service, finding that it had considered effects of a timber sale in the context of past and reasonably foreseeable logging; the agency had constructed mathematical models and had performed extensive field investigations to calibrate and verify its models. It had also actively sought public comment.[5]

6.4.4.2.2 Beginning the cumulative impact analysis To ensure the inclusion of resources that are most susceptible to degradation, cumulative impacts can be anticipated by considering where cumulative effects are likely to occur and what actions would most likely produce cumulative effects. In initiating a cumulative impact analysis, practitioners should:[6]

- Determine the area that would be affected.
- Make a list of the resources within that zone that could be affected by the proposed action.
- Determine the geographic areas occupied by those resources outside of the project impact zone.

6.4.4.2.3 Disregarding future actions Future actions can generally be disregarded if: [7]

- They lie outside the geographic boundaries or time frame established for the cumulative effects analysis.
- They will not affect resources that are the subject of the cumulative effects analysis.
- Their inclusion in the analysis is considered to be arbitrary (i.e., lacks a logical basis for inclusion).

The courts have struggled with the problem of determining when future actions can be disregarded as being "remote or speculative" versus those that must be analyzed. In one case, a court concluded that an EA prepared for mining operations did not need to consider the cumulative impact of other planned mines. This decision was premised on the fact that there was no practical commitment to future mining operations. The court concluded that an NEPA analysis must generally consider impacts of other proposals "... only if the projects are so interdependent that it would be unwise or irrational to complete one without the others."[8]

6.4.4.2.4 Evaluating cause-and-effect relationships Determining how a particular resource responds to an environmental disturbance is essential in determining the cumulative effect of multiple actions. Analysts must therefore gather information about cause-and-effect relationships. Once all the important cause-and-effect pathways are identified, the analyst determines how the environmental resource responds to a potential disturbance. Cause-and-effect relationships for each resource are used in computing the cumulative effect resulting from all actions that will be evaluated.

Typically, analysts will determine the separate effects of the proposed action, and other past, present, and reasonably foreseeable future actions. The cumulative effect can then be summated once each group of effects is determined.

6.4.4.2.5 Performing a qualitative analysis A cumulative impact analysis is sometimes limited to a qualitative evaluation because the cause-and-effect relationships are poorly understood. In still other cases, there may not be sufficient site-specific data available to permit a quantitative analysis. Faced with such constraints, analysts might want to consider performing a qualitative analysis in a manner similar to that shown in the following example (Table 6.4). If no numbers are available, the analyst may categorize the magnitude of effects using qualitative descriptors such as "high," "medium," or "low."

6.4.4.2.6 Considering related actions In one case, an agency was sued for preparing individual EAs on separate mining claims that involved a cumulatively significant impact. The court concluded that an EIS was necessary when a number of related actions cumulatively have a significant environmental impact, even if the separate actions, by themselves, would not. In the words of the court, "...once the cumulative impact of a number of mining claims crosses the threshold of [a] significant effect on the environment, a discussion of those cumulative effects in individual EAs no longer complies with NEPA."[9] Early preliminary consideration of possible related actions that might trigger an EIS due to potentially significant cumulative impacts is advisable before delving into the process of writing an EA.

6.4.4.2.7 Actions on private lands Case law indicates that impacts from unrelated actions on private lands must still be considered in cumulative impact analysis, not just impacts from actions of federal lands. In one case, the Eighth Circuit Court ruled that an EA must consider the impacts of activities reasonably expected to occur on private lands.[10] Rarely should the cumulative impact analysis in an EA ever require field studies or other investigative activities involving entry onto private lands other than private lands directly involved in the proposed federal action. Brief review of readily available published data sources should be more than adequate. Detailed site-specific environmental reports are generally not available for most private lands; available data is generally limited to broad-scale mapping sources such as topographic maps, National Wetland Inventory maps, geological maps, and zoning and land cover maps, as well as searchable databases maintained by some federal and state agencies.

6.4.4.2.8 Considering connected actions In 1988, the U.S. Forest Service was challenged for preparing nine separate EAs on connected actions. In reviewing the case, the court found that the plaintiffs had raised serious questions as to whether these timber sales would result in a cumulatively significant impact. The court found that the agency's FONSIs were

Chapter six: Environmental impact assessment

Table 6.4 Example of a cumulative effects analysis using quantitative descriptions

Environmental resources	Past actions	Present actions	Proposed actions	Future actions	Cumulative impacts
Vegetation	30% of presettlement vegetation lost	1% of vegetation lost this year	3% of existing vegetation would be lost	1% of vegetation lost yearly for next 15 years	49% of presettlement vegetation lost over next 15 years
Wetlands	30% of presettlement vegetation lost	1% of wetlands lost this year	9% of existing wetlands would be lost	3% of wetlands lost yearly for next 15 years	85% of presettlement wetlands lost over next 15 years
Turtles	20% of presettlement turtles lost	2% of turtles lost this year	6% of existing turtles would be lost	3% of turtles lost annually for next 15 years	73% of presettlement turtle population lost over next 15 years

inappropriate because the EAs did not adequately address connected actions and the cumulative effects of proposed and contemplated actions. The court concluded that the scope of these connected actions were broad enough so as to require preparation of an EIS.[11]

In the same year, another proposed action was challenged on similar grounds. The EA failed to evaluate the cumulative effects of connected actions involving reconstruction of a 17-mile segment of a 70-mile road, as well as other segments of the road reconstruction project, related timber sales that justified the entire project, and other reasonably foreseeable future actions. The court found that the connected actions, in addition to other reasonably foreseeable future actions, could result in a cumulatively significant impact. This was because there was an inextricable nexus between the logging operations and the road construction. The court concluded that the EA failed to evaluate the ongoing and future timber harvest and the road reconstruction.[12]

One of the most important cumulative impact cases involved the U.S. Fish and Wildlife Service. The agency had prepared other independent documents indicating that related and cumulative impacts might be leading to aquatic habitat degradation. Such degradation was unaccounted for in the individual EAs that it had prepared. The court found that lack of an overall effort to evaluate cumulative impacts could result in detrimental effects on the recovery of the wolf population. This was sufficient to raise serious questions regarding whether the road and the timber sales would result in a significant cumulative impact. The agency was ordered to prepare an EIS to analyze such effects.[13]

6.4.4.2.9 Differences in cumulative impact analyses between EAs and EISs An agency bears the burden of proof for demonstrating in an EA and FONSI that no significant impacts, including cumulative impacts, would result from a proposed action. In determining whether an EIS needs to be prepared, the analysis of cumulative effects in an EA may sometimes require even more rigor than is generally called for in an EIS. However, a brief, perfunctory analysis of cumulative impacts is normally adequate for most simple EAs.

6.4.4.2.10 The Fifth Circuit Court decision The Fifth Circuit Court concluded that the question of determining when an EIS is required (the objective of an EA) may necessitate a broader analysis of cumulative impacts than is generally necessary in an EIS. According to the court, an EA "should consider (1) past and present actions without regard to whether they themselves triggered NEPA responsibilities, and (2) future actions that are reasonably foreseeable, even if they are not yet proposals and may never trigger NEPA-review requirements."[14]

Specifically, the Fifth Circuit found that:

> ... although cumulative impacts may sometimes demand the preparation of a comprehensive EIS, only the impacts of proposed, as distinguished from contemplated, actions need be considered in scoping an EIS. In a case like this one, on the other hand, where an EA constitutes the only environmental review undertaken thus far, the cumulative impacts analysis plays a different role. ...
>
> (The NEPA Regulations) require an analysis, when making the NEPA-threshold decision, (to determine if) it is reasonable to anticipate cumulatively significant impacts from the specific impacts of the proposed project. ... (W)hen deciding the potential significance of a single proposed action ... a broader analysis of cumulative impactions that are not yet proposals and from actions—past, present, or future—that are not themselves subject to the requirements of NEPA.

The court cautioned that it did not mean to imply that "... consideration of cumulative impacts at the threshold stage will necessarily involve extensive study or analysis of the impacts of other actions." Instead, the court emphasized that the EA analysis should be limited to determining whether "... the specific proposal under consideration may have a significant impact."

The court went on to state that, at a minimum, an EA must demonstrate that the agency considered impacts from "past, present, and reasonably foreseeable future actions regardless of what agency, (federal or non-federal), or person undertakes such other actions." According to the court, the extent of the analysis depends on the scope of the affected area and the extent of other past, present, and future activities.

6.4.5 General recommendations

Several general recommendations for describing environmental impacts in an EA (or EIS) are detailed in Table 6.5. Although the list may appear long and complicated, it outlines what is essentially a thought process for describing impacts in NEPA documents. A brief evaluation following that thought process is usually adequate for simple EAs. A sliding-scale approach applies to the effort normally needed for more complex EAs.

Different means for evaluating potential impacts may be identified through comments gathered during EA scoping, if performed, or through

Table 6.5 Recommendations for describing environmental impacts

- Identify, but do not address, clearly insignificant impacts in detail. Provide only sufficient information to demonstrate that greater consideration is not needed.
- Address environmental impacts in proportion to their potential significance. Focus the analysis on issues that may have significant impacts.
- Identify all potential nontrivial impacts, including those that may not be the primary focus of the alternative.
- Quantify impacts to the extent practicable, consistent with the sliding-scale approach and taking into account available project information and design data.
- Describe the likelihood of potential impacts whenever possible.
- Do not confuse the description of impacts from the no-action alternative with the description of the affected environment. For example: The affected environment's water quality discussion might describe the general hydrological characteristics and contaminate rates. Also, this discussion would identify, as appropriate, existing water quality permits and specify the status for contaminant pollutants. In contrast, impact assessment for the no action alternative would include projections of future effluents rates without the proposed action.
- Identify possible indirect impacts, and indicate the degree to which these impacts are uncertain. For example describe the installation of a flood control dam that might *directly* interrupt fish migration and spawning; over time, birds (which feed on salmon) might *indirectly* suffer from a loss of food supply.
- Identify any responsible opposing views regarding how to conduct impact analysis or interpret conclusions.
- Note that compliance with requirements does not necessarily imply that there is no potential for significant impacts (demonstrating compliance does not substitute for analyzing impacts). Consider the quality of the affected environment. An action that would contaminate a pristine area or increase pollutant levels close to applicable limits may have significant impacts.
- Evaluate impacts for as long as they are reasonably foreseeable (not speculative), including foreseeable long-term as well as short-term effects.
- Consider impacts within geographic boundaries appropriate for each resource reviewed.
- Consider presenting the impacts by resource area (rather than by alternative) to allow for direct comparison among alternatives.
- Separately address effects from all activities encompassed within each alternative (e.g., construction, operation, waste management, transportation, decontamination, and decommissioning).

(continued)

Table 6.5 Recommendations for describing environmental impacts (continued)

- Where environmental standards or requirements are directly relevant to limiting impacts, identify these requirements, describe their conditions (e.g., release limits), and note whether the alternative threatens a violation of any applicable environmental requirement.
- In addition to identifying pollutants that would be released and wastes that would be produced, identify potential effects from these substances (e.g., human diseases, and effects on plant and animal populations and ecosystem functions). A quantified release rate should not be the endpoint in impact analysis.
- Acknowledge uncertainty and incompleteness in the data. Where the uncertainty is significant or a major factor in understanding the impacts, explain how the uncertainty affects the analysis.
- Do not attempt to quantify impacts on environmental resources when it is clear from the context that impacts would be virtually absent. Provide a brief negative declaration, such as, "The project would not affect threatened or endangered species or their habitats," and provide appropriate references, consultation letters, or explanation. Compare environmental impacts in their appropriate context. Do not use regional, national, or global comparisons that might trivialize the significance of a local impact.
- Understand that relative comparisons do not provide absolute information, and both relative and absolute information generally are needed.

other information. The EA should (1) acknowledge the controversy (i.e., the differences of opinion or fact) and (2) explain the basis for choice of methodology in a manner that demonstrates that the analysis in the EA is technically sound and provides a sufficient basis for decision making. EA authors should consider explaining how use of the different methodology would affect the conclusions, as this different perspective might be useful to the decision making process. In some cases, it may be prudent and useful to present the results of using the alternative data, assumptions, or methodologies.

6.4.5.1 Segmentation

With respect to segmentation claims, the courts affirmed that connected actions are those that are automatically triggered by another action or are not independently justified. One recent NEPA case concerned a U.S. Army Corps of Engineers (Corps) connector road project with three phases. The ruling in this case was that the Corps did not segment these actions as phase II had "independent utility" and would not automatically trigger phase III.* A common segmentation controversy arises concerning EAs and EISs that address development of major industrial facilities such as

---

* *Northwest Bypass Group v. U.S. Army Corps of Engineers* (D. N.H., April 22, 2008; Case No.: 06-00258)

power plants. Agencies often seek evaluation of "single and complete projects" that might include not only the power plant or other main facility but also supporting appurtenances such as access roads, pipelines, utility lines, and transmission lines. Design and engineering of the appurtenances often lag behind that of the main facility and may not yet be sufficiently advanced to allow assessment of impacts at the same time that the EA or EIS is written for the main facility. It is sometimes possible to argue that some appurtenances such as a pipeline or transmission line might have "independent utility" if it can be demonstrated that the appurtenance would likely be constructed to serve other existing and proposed facilities even if the subject facility were never constructed. Other times, opponents of the action might argue that were the main facility never built and operated (i.e., "but for the main facility), the appurtenances could never be justified.

6.5 Human health effects

Exposure and dose are neither health effects nor environmental impacts. Rather, they cause the health effects. Human health effects from exposure to chemicals may be both toxic effects (such as nervous system disorders) and cancer. The potential human health effect resulting from exposure to low doses of radiation is principally that of cancer.

A principal problem commonly encountered in estimating effects from human exposure to chemicals or radiation is the failure (or inability in some cases) to carry the analysis to completion, i.e., to identify, and quantify when appropriate, potentially significant health effects (e.g., number of deaths).

It is appropriate, but not at all sufficient for purposes of analysis, to state that workers would have to comply with all applicable requirements, and that exposure to workers and the public would be minimized by using appropriate and approved safeguards and procedures, or that exposure to workers and the public would be maintained as low as reasonably achievable below applicable dose limits. However, this does not represent necessarily proof that the impact is insignificant.

In general, health effects should be investigated for both routine operations and accident scenarios. Provide the basis for health effects calculations, as it may be misleading to present only the resulting estimates. As necessary, present dose, or dose-to-risk (health effects) conversion factor, potential health effects calculated for the year of maximum dose and for the total period of estimated exposure, and any other germane information.

The term "excess fatal cancers" refers to those latent cancer fatalities beyond what would be expected to occur in the population absent the radiation exposure. When using the term "risk" it is that "risk" be defined and that the context for its use is provided. If "risk" or "probability" is used in describing potential effects, be certain to state the effect the probability describes (e.g., the probability of cancer death, probability of high dose rate, probability of a particular accident scenario). Present the probability and potential effect in addition to any risk estimate.

A fatal cancer would not be expected among this population. However, it would not be appropriate to interpret this result as indicating that there is zero risk of harm to workers from the radiation exposure.

6.5.1 Recommendations

Recommendations for evaluating human health effects are detailed in Table 6.6.

6.6 Accident analyses

An NEPA analysis might need to address the effects of a potential accident so as to inform the decision maker and public about reasonably foreseeable adverse consequences associated with the proposal. The term "reasonably foreseeable" extends to events that might have catastrophic consequences, even if their probability of occurrence is low, provided that the analysis of the impacts is supported by credible scientific evidence, is not based on pure conjecture, and is within the rule of reason (§1502.22).

Table 6.6 Recommendations for evaluating human health effects

- Establish a defensible period of estimated exposure by how long an alternative would expose workers or the general public. In general, impacts should be analyzed for as long as they are reasonably expected to occur.
- In general, it is considered good practice to provide estimates of potential health effects from an exposure for three subsets of populations and maximally exposed individuals in those populations:
  (1) Involved workers (participants at the location of the action)
  (2) Noninvolved workers (workers that would be on the site of the alternative but not involved in the action)
  (3) Members of the general public
- Apply the sliding-scale approach when characterizing human health effects.
- Analyses generally should be based on realistic exposure conditions. Where conservative assumptions (i.e., those that tend to overstate the impact) are used, describe the degree of conservatism, and characterize the "average" or "probable" exposure conditions if possible.
- Consider all potential routes of exposure, not just the most obvious route.
- Where the proposed activities might result in the air suspension of contaminated soils, consider the downwind exposure of the public to suspended particles.

With the exception of 40 *Code of Federal Regulations* §1502.22, the CEQ has not issued detailed guidance for addressing accident analyses. This section is intended to fill this gap, as it provides guidance for performing an accident analysis that is generally applicable to the preparation of both EAs and EISs.[15]

For the purposes of NEPA, an accident can be viewed as an unplanned event or sequence of events that results in undesirable consequences. Accidents can be caused by equipment malfunction, human error, or natural phenomena.

6.6.1 *Sliding scale*

Consistent with the principle that impacts be discussed in proportion to their significance (§1502.2[b]), analysts should use a sliding-scale approach in determining whether an accident analysis is appropriate, as well as the degree of effort that should be expended in performing such an analysis. Practitioners must apply professional judgment in determining the appropriate scope and analytical requirements. For example, practitioners need to determine the appropriate range and number of accident scenarios to consider, the level of analytical detail, and degree of conservatism that should be applied. A sliding-scale approach is particularly useful in making these determinations (Table 6.7).

The term "risk" used in the second bullet in Table 6.7 can be used to express the general concept that an adverse effect could occur. However, in quantitative assessments, it is more commonly understood to refer to the numeric product of the probability and consequences. "Risk" is used in the latter way in this text.

6.6.2 *Overview*

An accident is an event or sequence of events that is not intended to happen, and indeed might not happen during the course of operations. The probability that a given accident will occur within a given time frame,

Table 6.7 Key factors to consider in applying a sliding-scale approach to an accident analysis

- Severity of the potential accident impacts in terms of the estimated consequences
- Probability of occurrence and overall risk
- Context of the proposal
- Degree of uncertainty regarding the analyses
- Level of technical controversy regarding the potential impacts of the proposal

however, can be estimated. The probability of occurrence is expressed by a number between 0 (no chance of occurring) and 1 (virtually certain to occur). Alternatively, instead of a probability of occurrence, one can specify the frequency of occurrence (e.g., once in 200 years, which also can be expressed as 0.005 times per year).

An accident scenario is the sequence of events, starting with an initiator, that make up the accident. It is important to distinguish the probability (or frequency) of the accident initiator from that of the entire scenario; the latter quantity is of primary interest in NEPA accident analyses as it expresses the chance (or rate) that the environmental consequences will occur.

As used in this chapter, the environmental consequences of an accident are the effects on human health and the environment. In discussing an accident's effects on human health, it is both conventional and adequately informative to consider three categories of people: involved workers, noninvolved workers, and the maximally exposed individuals in these categories and the collective harm to each population; for example, this might involve identifying and quantifying, as appropriate, potential health effects (e.g., number of latent cancer fatalities).

In the context of analyzing accidents, the environment includes biota and environmental media, such as land and water, which can become contaminated as the result of an accident. The following guidance refers to effects on biota as ecological effects.

6.6.3 Accident scenarios and probabilities

This section provides guidance on addressing accident scenarios and probabilities.

6.6.3.1 Range of accident scenarios

Development of realistic accident scenarios that address a reasonable range of event probabilities and consequences is the key to an informative accident analysis. The set of accident scenarios considered should serve to inform the decision maker and public of the overall accident risks associated with a proposal. As appropriate, accident scenarios should represent the range, or "spectrum," of reasonably foreseeable accidents, which can include both low-probability/high-consequence accidents and higher-probability/(usually) lower-consequence accidents.

Remember that the purpose for preparing an accident analysis in an EA is different from that of an EIS. In an EA, the purpose of the accident analysis is to determine whether a significant impact could result, requiring preparation of an EIS. Thus, the EA analysis normally focuses on the accident that could result in the maximum reasonable consequences. In contrast, an EIS seeks to explore a range of different accidents and consequences that will assist the decision maker in choosing among various alternatives.

Thus where there is a potential for significant consequences, an EA normally focuses on the maximum reasonably foreseeable accident(s) that represent potential scenarios at the high-consequence end of the spectrum. A maximum reasonably foreseeable accident is usually an accident with the most severe consequences that can reasonably be expected to occur for a given proposal. Such accidents usually have very low probabilities of occurrence. Note that a maximum reasonably foreseeable accident is not the same as a "worst-case" accident, which almost always includes scenarios so remote or speculative that they are not reasonably foreseeable. Analysis of worst-case accidents is not required under NEPA.

An accident analysis does not necessarily end here. Accidents in the middle of the spectrum might also need to be evaluated, as they often contribute to or even dominate the overall accident risks. An exception to this guidance might involve circumstances where the consequences of the maximum reasonably foreseeable accident are small. In that case, analyzing only the maximum reasonably foreseeable accident would provide sufficient information regarding the overall accident risks of the proposal.

Equally important, a "bounding" approach that considers only the maximum reasonably foreseeable accident might not adequately represent the overall accident risks associated with the proposal. Further, bounding analyses might not enable a reasoned choice among alternatives and appropriate consideration of mitigation, because they tend to mask real differences among the alternatives.

6.6.3.2 Scenario probabilities

Accident scenarios can involve a series of events for which an initiating event is postulated. The initiating event would be followed by a sequence of other events or circumstances that result in adverse consequences. If these secondary events always occur when the initiator occurs (i.e., the secondary events have a probability of one given that the initiator occurs), then the probability (or frequency) of the entire accident scenario is that of the initiator. Otherwise, the scenario probability would be the product of the conditional probabilities of the individual events.

Scenarios based on pure conjecture are to be avoided (§1502.22). Exercise good judgment in compounding conservatisms—evaluating a scenario by using multiple conservative values of parameters can yield unrealistic results. For example, in air dispersion modeling, it is nearly always unrealistic to assume only extremely unfavorable meteorologic conditions (i.e., stable or 95% most unfavorable conditions). In many cases, it would be appropriate to estimate and present accident consequences based on both neutral (50%, such as Pasquill Stability Class D) and unusually stable (95%, such as Pasquill Stability Class F) conditions. It is generally inappropriate, however, to assume only the most severe conditions for an otherwise appropriate and credible accident scenario and then fail to analyze the scenario because, by taking into account the lower probability associated with the stable atmospheric conditions, the overall probability is judged to be not reasonably foreseeable. Similarly, using estimates of plume centerline concentrations might be appropriate for evaluating impacts to maximally exposed individuals, whereas average plume concentrations would yield more realistic results for population impacts.

6.6.4 Risk

It is generally insufficient to simply present the reader with the "risk" of an accident (calculated by multiplying the probability of occurrence times the consequence). Presenting only the product of these two factors masks their individual magnitudes. Accordingly, risk should augment and not substitute for the presentations of both the probability of occurrence and the consequence of the accident.

6.6.4.1 Conservatisms

Practitioners must exercise professional judgment in determining the appropriate degree of conservatism to apply. Preparers should consider the fundamental purposes of the analysis (e.g., purpose of an EA versus that of an EIS), the degree of uncertainty regarding the proposal and its potential impacts (see further discussion of uncertainty below), and the degree of technical controversy. In short, accident analyses should be realistic enough to be informative and technically defensible.

6.6.5 Accident consequences

Guidance for addressing the consequences of an accident is provided in the following sections.

6.6.5.1 Involved and noninvolved workers

Noninvolved workers are those who would be within the vicinity of the proposed action, but not involved in the action. Any potential impacts to noninvolved workers should always be considered as part of an accident analysis.

Impacts to any involved workers should also be evaluated as part of an accident analysis. For example, fatal or serious nonfatal injuries might be expected because of a worker's close proximity to the accident. In some cases, credibly estimating exposures for involved workers can require more details about an accident than could reasonably be foreseen or meaningfully modeled. As a substitute, the effects can be described semiquantitatively or qualitatively, based on the likely number of people who would be involved and the general character of the accident scenario. For example, a qualitative analysis might indicate:

Seven workers would normally be stationed in the room where the accident could occur. While a few such workers might escape the room in time to avoid being seriously harmed, several would likely die within hours from exposure to toxic substances, and the exposed survivors might have permanent debilitating injuries, such as persistent shortness of breath.

A more detailed, semiquantitative discussion might be appropriate for analyzing proposals with substantially greater risks.

6.6.5.2 Uncertainty

A decision maker needs to understand the nature and extent of uncertainty in choosing among alternatives and considering potential mitigation measures. Thus, where uncertainties preclude quantitative analysis, the unavailability of relevant information should be explicitly acknowledged. The NEPA document should describe the analysis that is used, and the effect that the incomplete or unavailable information has on the ability to estimate the probabilities or consequences of reasonably foreseeable accidents (§1502.22).

Based on the prevailing circumstances, practitioners can compensate for analytical uncertainty by using conservative or "bounding" approaches that tend to overestimate potential impacts (see the earlier section that discusses some of the pitfalls of such approaches). In other circumstances, such as where substantial uncertainty exists regarding the validity of estimates, a qualitative description may suffice. In all cases, however, the NEPA document should explain the nature and relevance of the uncertainty.

Regardless of whether a qualitative or quantitative analysis is performed, references supporting scenario probabilities, and other data and assumptions used in the accident analysis, should be provided.

6.6.6 Intentionally destructive acts

An NEPA document might need to address potential environmental impacts that could result from intentionally destructive acts (i.e., acts of sabotage or terrorism). Intentionally destructive acts are not accidents *per se*.

Analysis of such acts (fire, explosion, missile, or other impact force) poses a challenge because the potential number of scenarios is limitless and the likelihood of attack is unknowable. Nevertheless, the physical effects of such destructive acts are generally similar to, or "bounded" by, the effects of accidents. For this reason, where intentionally destructive acts are reasonably foreseeable, a qualitative or semiquantitative discussion of the potential consequences of intentionally destructive acts should be included in the accident analysis.

The following is an example of a qualitative discussion that might be appropriate for a hypothetical proposal involving a terrorist act against a truck transporting a chemical agent:

> Explosion of a bomb beneath the transportation truck or an attack by an armor-piercing weapon is possible. However, analysis shows that the consequences of such acts would be less than or equal to those associated with a maximum reasonably foreseeable transportation accident.

6.7 Considering other environmental requirements

The following describes some of the principal environmental laws, rules, regulations, and executive orders (EOs) that may be relevant to an action subject to NEPA analysis. CEQ regulations 40 CFR 1500.2 and 40 CFR 1502.25 identify related environmental laws, rules, regulations, and EOs to be integrated concurrently to the fullest extent possible in EAs and EISs. Brief explanations of how the NEPA process has complied with these legal requirements should be presented in a chapter of the EA or EIS. The breadth of relevant laws, rules, regulations, and EOs varies with the individual action addressed and, as the federal government is dynamic rather than static, is in a continual state of flux.

The following executive orders and statutes have websites and additional references to view for more information. This is not to be used as a definitive nor an encyclopedic source for information about these requirements or how to comply with them.

Just because the text below discusses several specific topics frequently considered in an EA or EIS, it is not intended as an exhaustive list. The range of topics considered in any EA or EIS must rely on individual evaluation of the action at hand by qualified professionals combined, of course, with comments received during scoping efforts.

6.7.1 Clean air conformity requirements

The agency must ensure that their actions conform to applicable federal, state, or tribal implementation plans for achieving National Ambient Air Quality Standards (NAAQS: 40 CFR Part 50). To conform, federal actions must not contribute to new violations of the standards, increase the frequency or severity of existing violations, or delay timely attainment of standards in the area of concern (e.g., a state or a smaller air quality region such as a cluster of counties).

The conformity analysis (40 CFR Part 93) and NEPA process should be conducted in parallel and integrated to the extent practicable. The conformity analysis may involve two phases. First is the conformity review process to determine whether the conformity regulations would apply to an alternative (i.e., whether a conformity determination is needed). Second is the conformity determination process to demonstrate how an alternative would conform to the applicable implementation plan.

6.7.1.1 Recommendations

Complete the conformity review process for all analyzed alternatives in EAs and EISs to facilitate the comparison of alternatives with respect to air quality issues. A conformity review normally is not needed for the no-action alternative; a conformity review may be needed, however, if activities associated with the no-action alternative have pollutant air emissions that have not been subject to a prior conformity review.

Conduct the conformity determination process (if needed) for only the preferred alternative. Consider preparing a draft conformity determination for each analyzed alternative in cases where doing so would increase flexibility in making a final decision subsequent to the EA or allow comparison of the full cost requirements of each alternative. Explanation: The extent of analysis needed for a conformity determination coupled with the potential need to negotiate binding mitigation measures or offsets with non-DOE entities normally makes it impractical to complete the conformity determination process for all alternatives. Moreover, the conformity regulations do not require that conformity determinations be conducted for all alternatives.

6.7.2 Floodplain and wetland environmental review requirements

Executive Orders 11988 Floodplain Management (May 24, 1977) and 11990 Protection of Wetlands (May 24, 1977) direct federal agencies to undertake various actions to protect floodplains and wetlands. Wetlands are defined by EPA in 40 CFR 230.21 and by the U.S. Army Corps of Engineers in 33 CFR 328.3. Both agencies use the same technical definition of wetlands, which is based on specific criteria for vegetation, soils, and hydrology. Floodplains are defined in terms of probability of flooding during a normal year—for example, lands with a 1 percent probability of flooding in any year and thus an expected flood occurrence of 1 flood per 100 years are defined as the "100-year floodplain." Most agencies limit their consideration to 100-year floodplains, although some agencies such as DOE also consider impacts from certain types of activities (e.g., storage of hazardous materials) to the 500-year floodplain (area with an expected flood occurrence of once per 500 years.)

Agencies differ in how they address wetland and floodplain impacts, but all must demonstrate that they have considered avoidance and minimization of wetland and floodplain impacts when planning construction in wetlands or floodplains. DOE requires preparing a floodplain or wetland assessment for any action proposed in a floodplain and new construction proposed in a wetland. DOE's regulations implementing these executive orders, Floodplain and Wetland Environmental Review Requirements (10 CFR Part 1022), require that any floodplain or wetland assessment normally be included in an EA or EIS, if one is being prepared (10 CFR 1022.13(b)). Most other agencies also address wetlands and floodplains in the text of EISs or EAs for projects affecting such areas.

A floodplain or wetland assessment for DOE includes a description of the alternatives, a discussion of its potential effects on the floodplain or wetland (including a discussion of floodplain or wetland values), and consideration of alternatives (10 CFR 1022.4). The outcome of a floodplain assessment is documented in a floodplain statement of findings, which may be incorporated into a finding of no significant impact, final EIS, or record of decision, as appropriate (10 CFR 1022.14). A wetland statement of findings may be similarly prepared for a wetland assessment but is not required.

Although wetlands and floodplains are commonly considered together, they are by no means identical. Both have very specific definitions. Many wetlands occur within floodplains, but many others occur outside of floodplains. Likewise, many areas inside floodplains are not wetlands, and many areas outside floodplains are wetlands.

As a result of recent court decisions, not all areas meeting the technical definition of wetlands based on hydrology, soil, and vegetation criteria necessarily are waters of the United States that fall under jurisdiction of Section 404 of the Clean Water Act. Areas meeting the technical criteria for delineation of wetlands but failing to meet jurisdictional criteria under the Clean Water Act are commonly termed "isolated wetlands" or "nonjurisdictional wetlands." However, such areas are still wetlands. Executive Order 11990 requires federal agencies to consider all wetlands when planning federal development activities, irrespective of additional restrictions imposed by the Clean Water Act. Nonfederal proponents of actions impacting nonjurisdictional wetlands may not have to seek permits for those actions under Section 404 of the Clean Water Act, but federal agencies must consider the effects of federal actions on all wetlands.

6.7.3 National Historic Preservation Act (16 U.S.C. 470)

The National Historic Preservation Act of 1966 (NHPA) has two major components that affect the responsibilities of federal agencies.

First, under Section 106 of the NHPA, federal agencies are to consider the effects of their actions on historic resources that are either eligible for listing or are listed on the National Register of Historic Places. Section 106 of the National Historic Preservation Act of 1966 (NHPA) requires federal agencies to "take into account" the effect of their projects on historical and archeological resources and to give the Advisory Council on Historic Preservation the opportunity to comment on such effects. Where both NEPA and the NHPA are applicable, draft EAs and EISs must integrate NHPA considerations along with other environmental impact analyses and studies (40 CFR 1502.25). Additional requirements regarding consultation with external parties and other aspects of integrating NHPA and NEPA are found in regulations implementing Section 106 (36 CFR Part 800, Subpart B).

Secondly, Section 110 of the NHPA requires federal agencies that own or control historic resources to consider historic preservation of historic resources as part of their management responsibilities.

6.7.4 Other related legal requirements

The following indicates some but not all of the executive orders and statutory requirements that may apply to preparation of an EA.

6.7.4.1 Executive order requirements

Executive Order 12114—Environmental Effects Abroad: EO 12114 extends the purpose of NEPA abroad by requiring federal agencies to consider the environmental effects of major federal actions outside of the United States.

Executive Order 12866—Regulatory Planning and Review: EO 12866 requires federal agencies to consider socioeconomic impacts during rulemaking.

Executive Order 12898—Environmental Justice: EO 12898 requires federal agencies to consider the impacts of their actions on minority and low-income populations. The objective of EO 12898 is to ensure that the burden of environmental impacts is not borne disproportionately by disadvantaged communities and individuals. Those communities and individuals often do not have the money or time to participate in environmental public participation activities such as NEPA or to hire experts or attorneys to represent their interests in environmental planning processes such as NEPA. Environmental justice is discussed further in Section 6.8.

Executive Order 13112—Invasive Species: EO 13112 requires federal agencies to use authorities to prevent introduction of invasive species, respond to and control invasions in a cost effective and environmentally sound manner, and to provide for restoration of native species and habitat conditions in ecosystems that have been invaded. Pests accidentally introduced into the United States in the nineteenth and early twentieth centuries such as the chestnut blight fungus, Dutch elm disease fungus, and gypsy moth have dramatically altered much of the North American landscape; more recently introduced pests such as the emerald ash beetle and Asian longhorned beetle threaten to kill vast numbers of deciduous trees in the forests of the eastern and central United States. Zebra mussels, an exotic mollusk species accidentally introduced to the Great Lakes in the ballast water of ships, have dramatically altered the aquatic ecology of the Great Lakes and many other northern waters. Chinese snakehead fish accidentally introduced from aquariums to the waters of Maryland threaten to compete with recreational fish species in the Potomac River, and Burmese pythons kept as pets in Miami have escaped and multiplied in the Everglades. Introduced weeds such as garlic mustard, Japanese stiltgrass, and bush honeysuckle species have displaced forest understory in vast areas of eastern deciduous forests. Brazilian pepper, introduced to south Florida as an ornamental landscape shrub, now infests the understory of huge swaths of upland and wetland forests in Florida. Many of the most serious agricultural weeds such as cheatgrass, crabgrass species, foxtail grass species, and even the ubiquitous dandelion have their origin as introduced invasive species.

Executive Order 13158—Marine Protected Areas: EO 13158 requires federal agencies to identify actions that affect natural or cultural resources that are within a marine protected area (MPA). It further requires federal agencies, in taking such actions, to avoid harm to the natural and cultural resources that are protected by an MPA.

6.7.4.2 Other statutory requirements

Some additional requirements that are commonly encountered are described below.

Administrative Procedure Act (5 U.S.C. Subchapter II §§551-559): The Administrative Procedure Act (APA) requires public disclosure on federal rulemaking efforts and other actions that have the effect of rulemaking. This requirement has a stepped process, similar to NEPA, where rules are published first in draft form. After the public has had an opportunity to submit comments on the proposed rule a final rule is published. The concept of an administrative record comes from the judicial review section of the APA. The scope of the APA is not limited to environmental issues.

Data Quality Act (Public Law 106-554 § 515; H.R. 5658): The Data Quality Act requires the director of the Office of Management and Budget to issue guidelines to federal agencies regarding the assurance of quality, objectivity, utility, and integrity in information (including statistical information) disseminated by federal agencies.

Coastal Zone Management Act—Federal Consistency (16 U.S.C. 1451-1465): The Coastal Zone Management Act (CZMA) requires that federal actions that will have reasonably foreseeable effects on the land or water uses or natural resources of a state's coastal zone must be consistent with federally approved State Coastal Management Programs. This generally involves conducting consultation with affected State Coastal Management Programs. Each state identifies its coastal zone. States abutting the Great Lakes have coastal zones as do states abutting the tidal waters of the Atlantic and Pacific Oceans and Gulf of Mexico.

Endangered Species Act Section 7 (16 U.S.C. 1531-1544, 87 Stat. 884): Section 7 of the Endangered Species Act reads as follows:

> If an agency proposes, permits, or funds an action that may affect ESA listed species, it must initiate a Section 7 consultation as required by the Endangered Species Act. Staff responsible for ensuring NEPA compliance may be involved in the section 7 consultation. Sections of a NEPA document, such as information on the affected environment, may be used in consultation. Section 7 consultation must be completed before the FEIS is completed or the FONSI is signed.

Magnuson-Stevens Fishery Conservation and Management Act— Essential Fish Habitat (Public Law 94-265): Section 305(b)(2) of the amended Magnuson-Stevens Fishery Conservation and Management Act directs each federal agency to consult with the secretary with respect to any action authorized, funded, or undertaken, or proposed to be authorized, funded, or undertaken, by such agency that may adversely affect any essential fish habitat (EFH) identified under the Magnuson-Stevens Act. Implementing regulations for this requirement are at 50 CFR 600.

National Historic Preservation Act (16 U.S.C. 470): The National Historic Preservation Act (NHPA) has two major components that affect the responsibilities of federal agencies. First, under Section 106 of the NHPA federal agencies are to consider the effects of their actions on historic resources that are either eligible for listing or are listed on the National Register of Historic Places. Secondly, Section 110 of the NHPA requires federal agencies that own or control historic resources to consider historic preservation of historic resources as part of their management responsibilities.

National Marine Sanctuaries Act (16 U.S.C. 1431-1445): Section 304 (d) of the National Marine Sanctuaries Act requires federal agencies to engage the National Marine Sanctuaries Program (NMSP) in consultation whenever their actions are likely to destroy, cause the loss, or injure any sanctuary resource.

Note that impacts involving Native American (American Indian) lands, resources, and cultural sites are potentially subject to several statutes dealing with Native American tribes, such as the Native American Graves Protection and Repatriation Act.

6.8 Environmental justice

With respect to NEPA, the issue of environmental justice (EJ) began surfacing around 1994. Some of the principal direction and guidance documents that have since been issued are outlined below.

In February 1994, President Clinton signed Executive Order 12898, Federal Actions to Address Environmental Justice in Minority Populations and Low-Income Populations.[18] This executive order requires that each federal agency:

> make achieving environmental justice part of its mission by identifying and addressing, as appropriate, disproportionately high and adverse human health or environmental effects of its programs, policies, and activities on minority and low-income populations.

A presidential memorandum accompanying this executive order directs federal agencies to "... analyze the environmental effects ... of federal actions, including effects on minority communities and low-income communities, when such analysis is required by the National Environmental Policy Act."

Federal agencies are also instructed to "... provide opportunities for community input in the NEPA process, including identifying potential effects and mitigation measures in consultation with affected communities and improving the accessibility of meetings, crucial documents, and notices."

EJ has become a topic of special interest in recent years. Accordingly, this section provides practitioners with guidance on incorporating EJ considerations into both the preparation of EAs and EISs. This direction is based principally on a presidential executive order, and guidance developed by the President's Council of Environmental Quality, the U.S. Department of Energy, and the U.S. Environmental Protection Agency (EPA).[16]

The EPA defines EJ as:[17]

> The **fair treatment** and meaningful involvement of all people regardless of race, color, national origin, or income with respect to the development, implementation, and enforcement of environmental laws, regulations, and policies. Fair treatment means that no group of people, including racial, ethnic, or socioeconomic group, should bear a **disproportionate share of the negative environmental consequences** resulting from industrial, municipal, and commercial operations or the execution of federal, state, local, and tribal programs and policies" (emphasis added).

Implemented without prudence, EJ considerations can easily get out of hand, significantly adding to what might already be a costly and resource-intensive environmental process. Emphasis is therefore placed on providing the reader with a practical and balanced approach for addressing EJ.

To this end, a sliding-scale approach should be applied in determining the level of effort most appropriate for addressing EJ considerations; that is to say, tasks such as identifying populations, assessing impacts, and enhancing participation should be performed commensurate with the potential for sustaining disproportionately significant impacts.

Application of the sliding-scale approach:

- Helps ensure that impacts on minority and low-income populations are not overlooked.
- May be qualitative or quantitative.
- May be more or less detailed than analyses for impacts on the general population, depending upon the significance of impacts on minority or low-income populations.

6.8.1 General guidance

The following describes two principal sources of guidance on performing an assessment of EJ.

6.8.1.1 CEQ guidance

In 1997, the CEQ issued EJ guidance under the National Environmental Policy Act. This guidance document provides direction on incorporating EJ into the NEPA process.[19] This guidance states that the presidential executive order signed in 1994 does not change prevailing legal thresholds and statutory interpretations under NEPA and existing case law. However, it emphasizes that agency consideration of impacts on minority or low-income populations can identify disproportionately high and adverse impacts that are significant and that might otherwise be overlooked.

The guidance goes on to point out that environmental justice issues encompass a broad range of impacts covered by NEPA, including impacts on the natural or physical environment and related social, cultural, and economic impacts. This guidance acknowledges that environmental justice issues can arise at any step of the NEPA process, and agencies should consider these issues at each and every step of the process, as appropriate.

This guidance states that environmental impacts to minority and low-income populations do not have a different threshold for "significance" from other impacts, but specific consideration of impacts on minority and low-income populations can identify "disproportionately high and adverse human health or environmental effects that are significant and that otherwise would be overlooked."

6.8.1.2 Environmental Protection Agency guidance

In 1998, EPA issued Guidance for Incorporating Environmental Justice Concerns in EPA's NEPA Compliance Analyses.[20] This guidance applies to NEPA reviews conducted by EPA.

Soon after, EPA issued EPA Guidance for Consideration of Environmental Justice in Clean Air Act Section 309 Reviews.[21] This guidance applies to EPA reviews (under Section 309 of the Clean Air Act) of EISs prepared by other federal agencies.

6.8.1.3 Department of Energy guidance

As of this writing, the U.S. Department of Energy has issued draft guidance on incorporating EJ into its NEPA process. The document concentrates on how to assesses impacts and how to enhance public participation among minority and low-income groups.[22]

6.8.2 Determining the appropriate level of NEPA review

Environmental impacts to minority and low-income populations can affect the level of required NEPA review. Specifically, they can affect whether an action can be implemented under a categorical exclusion or EA.

6.8.2.1 Categorical exclusions

Before an agency can categorically exclude a proposed action from NEPA review, it must determine, among other things, that there are no extraordinary circumstances related to the proposal that might affect the significance of its environmental effects. Extraordinary circumstances represent unique situations presented by specific proposals, such as scientific controversy about the proposal's environmental effects, uncertain effects, or unresolved conflicts concerning alternative uses of available resources. An extraordinary circumstance exists if a proposal that would normally be categorically excluded would have a disproportionately high and adverse affect on a minority or low-income population. In such cases, preparation of at least an EA would normally be required.

6.8.2.2 Environmental assessments

To issue a FONSI for a proposed action, the EA needs to demonstrate that there would be no disproportionately high and adverse impacts to minority or low-income populations. Alternatively, if such an impact should occur, the FONSI could commit to specific measures that would ensure that such an impact would be mitigated to the point of nonsignificance.

6.8.3 Analyzing environmental impacts

Agencies are instructed to evaluate proposals for their potential to produce disproportionately high and adverse human health or environmental effects.

Practitioners should note that there might be cultural differences among stakeholders regarding what constitutes an impact or the severity of an impact. For example, an Indian tribe might regard providing general public access to a particular mountain as desecration of its sacred site. Agency officials should also recognize that risk perceptions can vary widely, and commenters might disagree with the agency's underlying assumptions concerning risk factors.

6.8.4 Evaluating high and adverse impacts

On completing the analysis of impacts on the general population, analysts should determine, consistent with the CEQ guidance, whether any impacts on a minority or low-income population have the potential to be "disproportionately high and adverse."[23] A two-step approach is warranted. Specifically, practitioners should judge whether:

1. The impacts on a minority or low-income population would be potentially significant, within the meaning of NEPA (i.e., high and adverse).
2. Any potentially significant (i.e., high and adverse) impacts would disproportionately affect a minority or low-income population relative to the general population.

Impacts to a minority or low-income population that are considered to have a potential to be disproportionately high and adverse are analyzed. As with the analysis of impacts to the general population, a sliding-scale approach should be utilized in determining the level of assessment necessary to make judgments regarding the significance of impacts on minority and low-income populations. That is, one should perform a less rigorous analysis of proposals with a clearly small potential for impact, while devoting correspondingly more attention to actions where the potential for significance is greater.

Attention should focus on identifying and evaluating impacts to minority and low-income populations that would be different from the impact on the general population. A qualitative assessment can often be sufficient to provide the decision maker with information on which to base informed decisions. Where such differences are trivial, include only enough discussion to show why more study is not warranted.

Approach any investigation of impacts to minority or low-income populations as a subset of impacts on the general population. If appropriate, this should be done on a resource-by-resource basis (e.g., air quality, water quality) or impact area (e.g., health impacts, facility accidents, cumulative impacts). Any special mechanisms by which an impact could affect a minority or low-income population might also need to be described. The size of the population and its geographic area should be indicated.

Consider, as appropriate, whether the proposal would:

- Affect or deny access to any natural resource on which the minority or low-income population (but not the general population) depends for cultural, religious, or economic reasons (e.g., a plant from which art is made, and perhaps sold for profit).
- Affect a minority or low-income population's food source by reducing its abundance (e.g., development that would eliminate land habitat where food animals forage, or that would increase silt in a stream that is fished).

Practitioners should note that there might be cultural differences among stakeholders regarding what constitutes an impact or the severity of an impact. For example, an Indian tribe might regard providing general public access to a particular mountain as desecration of its sacred site. Agency officials should also recognize that risk perceptions can vary widely, and commenters might disagree with the agency's underlying assumptions concerning risk factors.

6.8.4.1 Factors used in determining whether an impact is disproportionately high and adverse

The CEQ guidance document, Environmental Justice Guidance under the National Environmental Policy Act, presents factors to consider when judging the importance of disproportionately high and adverse health and environmental impacts on minority and low-income populations (see Table 6.8).

It is important to note that economic or social effects are not by themselves considered significant (i.e., requiring preparation of an EIS). However, when an environmental impact statement is prepared, and economic or social and natural or physical environmental effects are interrelated, then the EIS will discuss all of these effects on the human environment (§1508.14). Based on this provision, it is reasonable to conclude that an EA does not need to consider EJ where the impacts are simply social and economic in nature.

Table 6.8 Factors useful in judging whether impacts are disproportionately high and adverse

Disproportionately high and adverse human health effects

When determining whether human health effects are disproportionately high and adverse, agencies are to consider the following three factors, to the extent practicable:

1. Whether the health effects, which might be measured in risks and rates, are significant (as employed by NEPA) or above generally accepted norms. Adverse health effects may include bodily impairment, infirmity, illness, or death.
2. Whether the risk or rate of hazard exposure by a minority population, low-income population, or Indian tribe to an environmental hazard is significant (with respect to NEPA) and appreciably exceeds or is likely to appreciably exceed the risk or rate to the general population or other appropriate comparison group.
3. Whether health effects occur in a minority population, low-income population, or Indian tribe affected by cumulative or multiple adverse exposures from environmental hazards.

Disproportionately high and adverse environmental effects

When determining whether environmental effects are disproportionately high and adverse, agencies are to consider the following three factors, to the extent practicable:

1. Whether there is or will be an impact on the natural or physical environment that significantly (with respect to NEPA) and adversely affects a minority population, low-income population, or Indian tribe. Such effects may include ecological, cultural, human health, economic, or social impacts on minority communities, low-income communities, or Indian tribes when those impacts are interrelated to impacts on the natural or physical environment.
2. Whether environmental effects are significant (with respect to NEPA) and are or may be having an adverse impact on minority populations, low-income populations, or Indian tribes that appreciably exceeds or is likely to appreciably exceed that on the general population or other appropriate comparison group.
3. Whether the environmental effects occur or would occur in a minority population, low-income population, or Indian tribe affected by cumulative or multiple adverse exposures from environmental hazards.

6.8.4.2 Focusing on impacts that would be different

The analysis should focus on identifying and evaluating impacts to minority and low-income populations that would be different from the impact on the general population. A qualitative assessment can often be sufficient to provide the decision maker with information on which to base informed decisions.

Table 6.9 Factors useful in identifying unique pathways, exposures, or cultural practices that might need to be considered

- Are exposure pathways or rates of exposure for minority and low-income populations different from exposure pathways or rates for the general population? Different pathways or rates could result from variations in:
  —physical location of a population's residences, workplaces, or schools.
  —dietary practices, such as:
- Consumption of wild plants, or subsistence hunting, fishing, or farming.
- Differential selection of foods that might have high concentrations of contaminants (for example, bottom-feeding fish or fish that feed on bottom-feeding organisms can bioconcentrate fat-soluble contaminants from sediments, and organ meats, such as elk liver, might have bioaccumulated such contaminants)
  —Water supplies, such as use of surface or well water for drinking or irrigation.
- Are there any known health, social, or economic conditions of a minority or low-income population that would result in a greater impact? For example, would there be a greater frequency of dose or greater impact from a dose, over a pathway shared with the general population? Such conditions could involve:
  —Different access to public services such as paved roads (unpaved roads increase exposure to contaminated fugitive dust).
  —Different access to health care (e.g., poor control of asthma can increase susceptibility to particulate matter in air).
- Does the minority or low-income population (but not the general population) use a natural resource or area for cultural, religious, or economic reasons? Such uses could include:
  —Plants for ceremonial or medicinal purposes, or from which art is made and perhaps sold for a profit.
  —Plant-gathering or clay-procurement areas.
  —Natural features mentioned in legends.
  —Ceremonial sites.

6.8.4.3 *Unique pathways, exposures, and cultural practice*

In assessing environmental impacts on minority and low-income populations, one should investigate the effects based on considerations such as special pathways, exposures, and cultural practices. Table 6.9 presents factors useful in identifying unique pathways, exposures, or cultural practices that might need to be considered. Table 6.10 provides a definition of the term "subsistence consumption," which is used in Table 6.9.

Table 6.10 Definitions of the term subsistence consumption

In its 1997 Environmental Justice Guidance under the National Environmental Policy Act, the CEQ issued guidance on key terms related to subsistence consumption as inserts in a reprinting of Executive Order 12898. The following guidance was developed by an Interagency Working Group on Environmental Justice, established by the executive order and chaired by the EPA.

Subsistence consumption of fish and wildlife—"Dependence by a minority population, low-income population, Indian tribe or subgroup of such populations on indigenous fish, vegetation, and/or wildlife, as the principal portion of their diet."

Differential patterns of subsistence consumption—"Differences in rates and/or patterns of subsistence consumption by minority populations, low-income populations, and Indian tribes as compared to rates and patterns of consumption of the general population."

6.8.4.4 Considering cumulative impacts

As appropriate, agencies should consider the potential for multiple or cumulative exposures.[24] The terms multiple or cumulative exposures are defined in Table 6.11.

In assessing cumulative impacts, practitioners need to recognize that minority and low-income populations might be affected by past, present, or reasonably foreseeable future actions in a manner different from that experienced by the general population.

Table 6.11 Definitions of terms multiple and cumulative environmental exposure

In its 1997 Environmental Justice Guidance under the National Environmental Policy Act, the CEQ presented definitions related to multiple and cumulative environmental exposure as inserts in a reprinting of Executive Order 12898. The following proposed definitions were developed by the Interagency Working Group on Environmental Justice, established by the executive order and chaired by the EPA.

Multiple environmental exposure—"Exposure to any combination of two or more chemical, biological, physical, or radiological agents (or two or more agents from two or more of these categories) from single or multiple sources that have the potential for deleterious effects to the environment and/or human health."

Cumulative environmental exposure—"Exposure to one or more chemical, biological, physical, or radiological agents across environmental media (e.g., air, water, soil) from single or multiple sources, over time in one or more locations, that have the potential for deleterious effects to the environment and/or human health."

6.8.5 Assessing significance and mitigation measures

On completing the analysis, the agency needs to determine whether potential impacts to minority and low-income populations are high and adverse, using the criteria specified for assessing significance in the CEQ NEPA regulations (§1508.27).

The presidential memorandum accompanying Executive Order 12898 states, "Mitigation measures outlined or analyzed in an environmental assessment, environmental impact statement, or record of decision, whenever feasible, should address significant and adverse environmental effects of proposed federal actions on minority communities and low-income communities."[25] Mitigation measures include steps to avoid, minimize, rectify, reduce, or eliminate the adverse impact (§1508.20).

The goal in mitigating disproportionately high and adverse effects is not to distribute the impacts proportionally or divert them to a non-minority or higher-income population. Instead, measures or alternatives should be developed to mitigate effects on both the general population and minority or low-income populations. In other words, the goal of mitigation is not to move the impacts around, but to identify practicable means to meet the purpose and need for taking action while avoiding or reducing undesirable environmental effects.[26]

6.8.6 CEQ guidance on mitigation and monitoring

A draft memorandum for public comment was issued by the CEQ in 2010. The draft memorandum provides guidance for departments and agencies of the federal government on the mitigation and monitoring of activities subject to the NEPA process.* The following summarizes this draft guidance.

6.8.6.1 Addressing mitigation in environmental assessments

When an agency develops and makes a commitment to implement mitigation measures to avoid, minimize, rectify, reduce, or compensate for significant environmental impacts,† then NEPA compliance can be accomplished with an EA coupled with a FONSI. Using mitigation to reduce potentially significant impacts to support a FONSI enables an agency to conclude the NEPA process, satisfy NEPA requirements, and proceed to implementation without preparing an EIS. In such cases, the basis for not preparing the EIS is the commitment to perform those mitigation measures identified as necessary to reduce the environmental impacts of the

---

* CEQ, Memorandum for Heads of Federal Departments and Agencies: Draft Guidance for NEPA Mitigation and Monitoring, February 18, 2010.
† 40 C.F.R. § 1508.20.

proposed action to a point or level where they are determined to no longer be significant. That commitment should be presented in the FONSI and any other decision document. CEQ recognizes the appropriateness, value, and efficacy of providing for mitigation to reduce the significance of environmental impacts; consequently, when that mitigation is available and the commitment to perform it is made, there is an adequate basis for a mitigated FONSI. This proposed draft guidance approves of the use of the "mitigated FONSI" when the NEPA process results in enforceable mitigation measures and thereby amends and supplements the previously issued CEQ guidance in the 1981 Questions and Answers About the NEPA Regulations.*

6.8.6.2 Ineffective mitigation measures

Mitigation commitments should be structured to include adaptive management in order to minimize the possibility of mitigation failure. However, if mitigation is not performed or does not mitigate the effects as intended by the design, the agency responsible should, based upon its expertise and judgment regarding any remaining federal action and its environmental consequences, consider whether taking supplementary action is necessary.[†] In cases involving an EA with a mitigated FONSI, an EIS may have to be developed if the unmitigated impact is significant. If an EIS is required, the agency must avoid actions that would have adverse environmental impacts or limit its choice of reasonable alternatives during the preparation of an EIS.[‡]

6.8.6.3 Environmental monitoring

Under NEPA, a federal agency has a continuing duty to gather and evaluate new information relevant to the environmental impact of its actions.[§] For agency decisions based on an EIS, the regulations require that "a monitoring and enforcement program shall be adopted ... where applicable for mitigation." In addition, the regulations state that agencies may "provide for monitoring to assure that their decisions are carried out and should do so in important cases."[¶] Monitoring plans and programs should be described or incorporated by reference in the agency decision documents.

The following are examples of factors that should be considered when prioritizing monitoring activities:

- Legal requirements from statutes, regulations, or permits.

---

* Commonly referred to as the "Forty Most Asked Questions."
† 40 CFR 1502.9(c).
‡ 40 CFR § 1506.1(a).
§ 42 U.S.C. § 4332(2)(A).
¶ 40 CFR § 1505.3.

- Protected resources (e.g., threatened or endangered species or historic site) and the proposed action's impacts on them. Degree of public interest in the resource or public debate over the effects of the proposed action and any reasonable mitigation alternatives on the resource.
- Level of intensity of impacts.

The form and method of monitoring can be informed by the agency's past monitoring plans and programs that tracked impacts on similar resources, and plans and programs used by other agencies or entities, particularly those with an interest in the resource being monitored. Monitoring methods include:

- Agency-specific environmental monitoring.
- Compliance assessment.
- Auditing systems.

Monitoring can be performed as:

- Part of a broader system for monitoring environmental performance.
- Stand-alone element of an agency's NEPA program.

6.8.6.4 Implementing mitigation measures

Consistent with the Open Government Directive, efficient systems for reporting should make use of existing agency websites to the maximum extent practicable.* The lead federal agency should ensure that responsible parties, mitigation requirements, and any appropriate enforcement clauses are included in documents such as authorizations, agreements, permits, or contracts. Such enforcement clauses, including appropriate penalty clauses, should be developed based on a review of the agency's statutory and regulatory authorities.

Other agencies that can provide direction and information useful in developing an effective monitoring program include: the U.S. Fish and Wildlife and National Marine Fisheries Services for evaluating potential impacts to threatened and endangered species; State Historic Preservation officers for evaluating potential impacts to historic structures; and the U.S. Army Corps of Engineers for evaluating potential wetlands impacts.

---

* OMB Memo Dec. 8, 2009 (available at http://www.whitehouse.gov/open/documents/open-government-directive).

6.8.6.5 Transparency in implementing mitigation measures and a monitoring program

NEPA requires all agencies of the federal government to make "information useful in restoring, maintaining, and enhancing the quality of the environment" (including information on mitigation monitoring of potentially significant adverse environmental effects) "available to states, counties, municipalities, institutions, and individuals."[*] It is the responsibility of the lead agency to make the results of relevant monitoring available to the public.[†] NEPA also incorporates the Freedom of Information Act (FOIA) by reference and ensures public access to documents reflecting mitigation monitoring and enforcement.[‡] The FOIA requires agencies to make available, through "computer telecommunications" (e.g., agency websites), releasable NEPA documents and monitoring results which, because of the nature of their subject matter, are likely to become the subject of FOIA requests.[§] Mitigation and monitoring reports, access to documents, and responses to public inquiries should be readily available to the public through online or print media, as opposed to being limited to requests made directly to the agency.

References

1. Eccleston, C.H., *The NEPA Planning Process: A Comprehensive Guide with Emphasis on Efficiency*, Section 12.4 and pp. 281–2, John Wiley & Sons Inc., New York, 1999; Eccleston, C.H., *Environmental Impact Statements: A Comprehensive Guide to Project and Strategic Planning*, John Wiley & Sons Inc., New York, 2000.
2. CEQ, *Considering Cumulative Effects Under the National Environmental Policy Act*, p. 4, January 1997.
3. EPA, *Consideration of Cumulative Impacts in EPA Review of NEPA Documents*, pp. 10–11, (EPA 315-R-99-002), May 1999.
4. *Fritiofson v. Alexander*, 772 F.2d 1225, 1243, 1245–6 (5th Cir 1985).
5. *Inland Empire Public Lands Council v. Schultz*, 992 F.2d 977, 982 (9th Cir 1993).
6. CEQ, *Considering Cumulative Effects Under the National Environmental Policy Act*, p. 15, January 1997.
7. CEQ, *Considering Cumulative Effects Under the National Environmental Policy Act*, p. 19, January 1997.
8. *Webb v. Gorsuch*, 699 F.2d 157, 161 (4th Cir 1983).
9. *Northern Alaska Environmental Center v. Lujan*, 15 ELR 21048 (D. Alaska, 1985).
10. *Sierra Club v. Forest Service*, 46 F.3d 835, 839 (8th Cir. 1995).
11. *Sierra Club v. U.S. Forest Service* (9th Cir. June 24, 1988).
12. *Save the Yaak Committee v. Block*, 840 F.2d 714 (9th Cir. 1988).

---

[*] 42 USC § 4332(2)(G).
[†] 40 CFR § 1505.3(d).
[‡] 42 USC § 4332(2)(C).
[§] 5 USC § 552(a)(2); 40 C.F.R. § 1506.6(f).

13. *Thomas v. Peterson*, 753 F.2d 754 (9th Cir. 1985).
14. *Fritiofson v. Alexander*, 772 F.2d 1225, 1243, 1245–1246, (5th Cir. 1985).
15. Department of Energy, *Analyzing Accidents Under NEPA*, draft guidance prepared by the Department of Energy's Office of NEPA Policy and Assistance, April 21, 2000.
16. Presidential Executive Order 12898, Federal Actions to Address Environmental Justice in Minority Populations and Low-Income Populations, signed by President Clinton in February 1994; CEQ Guidance, *Environmental Justice Guidance Under the National Environmental Policy Act*, December 1997; U.S. Environmental Protection Agency, *Guidance for Incorporating Environmental Justice Concerns in EPA's NEPA Compliance Analyses*, 1998; EPA Guidance for Consideration of Environmental Justice in Clean Air Act Section 309 Reviews, 1999; U.S. Department of Energy, *Draft Guidance on Incorporating Environmental Justice Considerations into the Department of Energy's National Environmental Policy Act Process*, April 2000.
17. U.S. Environmental Protection Agency, *Guidance for Incorporating Environmental Justice Concerns in EPA's NEPA Compliance Analyses*, 1998; *EPA Guidance for Consideration of Environmental Justice in Clean Air Act Section 309 Reviews*, 1999.
18. Presidential Executive Order 12898, Federal Actions to Address Environmental Justice in Minority Populations and Low-Income Populations, February 1994.
19. CEQ, *Environmental Justice Guidance Under the National Environmental Policy Act*, December 1997.
20. U.S. Environmental Protection Agency, *Guidance for Incorporating Environmental Justice Concerns in EPA's NEPA Compliance Analyses*, 1998.
21. U.S. Environmental Protection Agency, *EPA Guidance for Consideration of Environmental Justice in Clean Air Act Section 309 Reviews*, 1999.
22. U.S. Department of Energy, *Draft Guidance on Incorporating Environmental Justice Considerations into the Department of Energy's National Environmental Policy Act Process*, April 2000.
23. CEQ, *Environmental Justice Guidance Under the National Environmental Policy Act*, December 1997.
24. CEQ, *Environmental Justice Guidance Under the National Environmental Policy Act*, December 1997.
25. Presidential Executive Order 12898, Federal Actions to Address Environmental Justice in Minority Populations and Low-Income Populations, February 1994.
26. EPA, *Guidance for Incorporating Environmental Justice Concerns in EPA's NEPA Compliance Analysis*, 1998.

chapter seven

Writing the environmental assessment

> When I was younger, I could remember anything, whether it had happened or not; but my faculties are decaying now and soon I shall be so I cannot remember any but the things that never happened. It is sad to go to pieces like this but we all have to do it.
>
> —**Mark Twain**

7.1 Introduction

Whether prepared by agency staff or contractors, federal agencies are ultimately responsible for both the preparation and content of their National Environmental Policy Act (NEPA) documents. This chapter discusses the process of actually writing an environmental assessment (EA). This chapter is the core of this book; hence it is somewhat longer than many other chapters and restates some of the more important principles stated regarding EAs in other chapters, especially the analytical processes introduced in Chapter 6.

Because individual agencies have internal requirements and nuances, readers are encouraged to consult their agency's NEPA-specific implementing procedures and internal guidance for any requirements that may supplement the regulatory direction provided in the Council on Environmental Quality (CEQ) NEPA regulations. Contractors and license or permit applicants preparing EAs for use by an agency should first thoroughly familiarize themselves with that agency's NEPA practices. Contractors frequently assume that extensive experience preparing NEPA documents for other agencies translates into adequate knowledge of the practices of an agency they have never supported.

Additionally, contractors with extensive experience preparing environmental impact statements (EISs) for an agency should study that agency's EA practices. Applying EIS experience to preparing EAs is one source of excessively lengthy or complicated EAs.

Personal communication through a kickoff meeting and frequent regular coordination calls is the easiest and most direct way to gain the necessary familiarization with an agency's EA practices and expectations. Reading EAs recently completed by the agency and reviewing recent NEPA case law can provide valuable insight. Preparers should inquire as to whether the agency has a model EA or model EA outline to follow. Otherwise, it is recommended that writers prepare a detailed outline of the EA for internal review before proceeding to write. Each section of the outline should be annotated with general objections and approximate page length. Ideally, contractors should specify an outline in the proposal or task order negotiations to ensure that adequate time and resources are available to prepare the desired EA.

7.2   General direction for preparing an EA

The direction provided in Section 7.1 applies to preparation of the entire EA although this book uses the phrases "writing an EA" and "preparing an EA" as synonyms. Yet one can really consider the writing of an EA (or EIS) as just a subset of the total effort of preparing it. Writing the document only comes following a sequence of planning, research, consultation, and coordination. The CEQ regulations properly refer to the preparers of an EIS, not the authors. CEQ knew that the NEPA process was much more than mere writing.

7.2.1   Proposals and contracting

Unlike EISs, most EAs are prepared directly by agency staff. However, many larger or more complex EAs are prepared by contractors hired by the responsible agency. The success of these EAs depends in no small part on having a good working relationship between the agency and the contractor, and that can only happen when a good contract is in place.

Most federal contracts for professional services, including environmental consulting services, fall into one of two categories: fixed price and cost reimbursable. Fixed price contracts are simple: the contractor agrees to perform a defined scope of work for an agreed-upon payment (price). The agency pays the price once the scope of work is successfully completed, without consideration of how many hours the contractor devoted to the work or to the direct or indirect expenses incurred. Because the contractor cannot generally bill the agency for unforeseen costs incurred in completing the scope of work, the risk associated with completing the scope of work within the proposal budget is borne by the contractor. Some environmental consulting services whose levels of effort:

- Are relatively predictable are well-suited to fixed price contracts.
- Are routinely prepared in a short period of time.

- Can be reported in a short report document, especially if a widely used template is available.
- Involve only one or a few closely related technical disciplines.
- Involve a small number of professional staff.

Cost reimbursable contracts are considerably more complicated. The agency pays the contractor for all costs incurred in completing the scope of work plus an additional amount representing a profit, typically referred to by professional services firms as "fee." Costs include direct and indirect components. Direct costs are clearly attributable to performing a scope of work, such as labor costs, costs for supplies such as paper and field equipment, and costs for travel. In professional services firms, labor costs typically constitute salary divided by total annual work hours. Nonlabor costs such as for supplies and travel are typically referred to as "other direct costs." Indirect costs are those necessary to basic operation of the firm but not readily assignable to any individual project. Such costs are sometimes termed "overhead" but are more commonly referred to by environmental consulting firms as "general and administrative" costs.

Other than EAs for very small and simple projects, NEPA document preparation is almost always best performed on a cost reimbursable basis. For EAs, agencies usually hire contractors only to write the largest and most complex of EAs. Any estimate of the effort needed to go through the NEPA process, whether via an EIS or long EA, is subject to considerable uncertainty. Unanticipated issues and necessary analyses may only become apparent following scoping. Scoping can even identify additional alternatives requiring detailed consideration. Even if an agency opts not to hold public scoping, which is not required for EAs, consultation with agencies such as the Fish & Wildlife Service and State Historic Preservation Office can identify additional issues, analyses, and alternatives. Moreover, EAs and EISs rely on professional interpretation of results by the preparers. The extent to which professionals working for the client agency, cooperating agencies, and reviewing agencies, as well as members of the public, question the interpretations is highly unpredictable.

Nevertheless, many agencies ask contractors to submit fixed price bids to prepare NEPA documents, especially EAs. The requested price can encompass not only writing the EA, but holding public scoping, preparing a notice of availability for the agency to publish in the *Federal Register*, drafting regulatory consultation letters, responding to public comments, and even preparing technical background papers such as wetland delineations and biological assessments. Even experienced NEPA contractors can feel challenged when trying to estimate a fixed price for such a broad and unpredictable suite of services. Some specific advice for contractors responding to requests for proposals (RFPs) for EAs or EISs is as follows:

- Although the RFP may request only a single bottom-line price, or perhaps prices for specific milestones, break the scope of work into small subtasks and estimate hours for each. This approach will help prevent overlooking work elements when pricing. Roll up the estimated hours to generate the total(s) requested in the RFP only as a final step.
- Read EAs recently completed by the subject agency, ideally for similar projects, to get an idea of the length and detail expected. The easiest place to look is on the agency's website.
- Read EISs recently completed by the subject agency, ideally for similar projects, but remember that EISs and EAs serve differing purposes and generally involve differing organization and writing styles.
- Brief calls to a few relevant state or local regulatory agencies can reveal possible complications that might be encountered. You may merely want to express interest in the project, not reveal to the agencies that you are bidding on a project.
- Pay attention to the fine details in the RFP such as number of interim drafts, number of copies, requirements for color graphics, and requirements for professional technical editing. These needs can substantially affect price.
- Attend any prebid meetings, if offered. If not, ask questions of the contract officer by phone. Seek as much clarification as possible; the contract officer knows where to draw the line and not provide any unfair answers.
- Remember that EAs are almost always substantially cheaper documents than EISs, often by an order of magnitude or more. If your EA bid is anywhere near what you would expect to bid on an EIS, even for an unusually complicated EA, you probably have overbid. Review the RFP and your assumptions again and revise your bid.
- Remember that EAs often do not include public scoping meetings, that EAs rarely involve responding to more than a few dozen comments, and that the "final" EA often consists only of an addendum responding to comments and an errata sheet. Scrutinize the RFP carefully and ask questions if you are unclear whether the client agency expects formal public scoping efforts, an extensive comment response effort, or extensive postcomment revisions.
- Including some "slush" hours to use to respond to impromptu questions from agency staff and other small unplanned activities. However, bidders must include only limited "slush" hours or else the bid will not win. The reality is that environmental consultants have to adhere to the scope outlined in the RFP and be prepared to insist on supplementary payment for significant deviations.

Most importantly, EA and EIS preparation contractors should state their assumptions in the text of their bids. Clearly articulated assumptions are the only way to submit a competitive fixed price bid on an EA without taking on a great risk of performing uncompensated work. Many agency clients may not like the assumptions, but they do understand the need for them. Even though agencies are supposed to consider technical as well as cost factors when evaluating bids, the fact is that the lowest bid—whether accompanied by assumptions or not—usually has a substantial advantage over higher bids. Key assumptions to include for an EA project are

- Number of scoping comments requiring response, if any.
- Specific sources of supporting data to be provided by the client agency.
- Specific analyses to be performed or data to be collected by the bidder.
- Specific background papers to be prepared by the bidder (examples include biological assessments, wetland delineations, archaeological surveys, and historic building surveys).
- Specific regulatory consultations to be performed by the bidder (examples include consultations with the Fish & Wildlife Service, State Natural Heritage Program, and State Historic Preservation Officer).
- Maximum page length of EA and any in-scope background papers.
- Number of drafts of each deliverable, including any interim drafts.
- Maximum number of comments requiring responses.

Bids must of course indicate that each deliverable will be submitted by the due dates specified in the RFP but should also state key assumptions underlying those due dates. Bids should state that the proposed deliverable dates presume a specific notice-to-proceed date as well as specific dates for receipt of materials or information promised by the client. Bids should also build in assumptions regarding receipt of appointments and responses from key regulatory agencies; many state offices are understaffed and can be slow in responding to data or consultation requests. Bidding is even more complicated when cooperating agencies are involved.

7.2.2 Staffing

Once an agency facing preparation of an EA has decided utilize internal staff or supplement internal staff by issuing a contract, the agency then must staff the EA preparation project. The number of professionals assigned to prepare an EA can vary from one or two to a small interdisciplinary team. Rarely do EAs involve the large interdisciplinary teams typical of large EISs. The act of preparing an EA can be considered a "project"; as such, a project manager should be designated. The project management duties are very simple for a short, focused EA and increase dramatically as the EA becomes more complex, requiring larger staffs.

Simple EAs can be authored by a single individual, often one employed directed by the agency. That individual, who will also function as the project manager for preparing the EA, may be an environmental generalist or a specialist in the environmental resource most central to the analysis. As examples, a biologist might author an EA on vegetation control or wildlife management, or a hydrologist might author an EA on a proposed water allocation project. If the agency lacks an appropriate specialist, it may direct an environmental generalist to write the EA. The generalist may have to seek technical assistance from other professionals, such as engineers, from within the ranks of the agency. Alternatively, the agency might elect to hire an environmental consulting firm with key technical competencies to write the EA.

Although few EAs address in detail the breadth of topics commonly covered in larger EISs, many of the larger EAs must cover multiple, often interrelated scientific fields of expertise. For example, an EA for a small construction project might have to include a detailed consideration of possible effects on biological and cultural resources as well as traffic impacts. Participation by a biologist, archaeologist, and traffic planner might be essential. The agency might task the three specialists to perform the three detailed analyses plus a generalist to provide direction and oversight (i.e., serve as a project manager), pull materials written by the specialists into the EA, and write the remainder of the EA, including brief text addressing minor issues. An even better approach would be for one of the specialists to also serve as the project manager, as long as the specialist also possessed a broad knowledge of NEPA.

Involving staff with expertise only in minor or peripheral subject areas is not a good idea and is one reason for excessively lengthy or otherwise poorly prepared EAs. Obviously, there is no need to involve an archaeologist in an EA for renovating the interior of an office building, even if the office building is historic and the skills of a historian may be called for. For instance, involving an archaeologist might result in an EA that includes a detailed description of the surrounding archaeology and a statement that the action would not disturb intact archaeological profiles. Less obvious is whether a biologist or traffic planner needs to be involved. There may be a need to discuss use of insecticides in the building or the possible effects of placing construction trailers under mature trees adjoining the building. There may be a need to discuss how moving construction equipment over local roads might affect traffic flow. However, it might be possible for a single generalist to succinctly but adequately discuss these issues. The generalist might be able to quickly confer with a biologist or traffic planner rather than having the specialists each write individual text for the EA.

7.2.3 Research

EA preparers should commence research early. Even if the agency plans to perform a formal scoping process, preparers should begin compiling information needed to perform the impact evaluation as early as possible. Many EAs (and EISs) use technical literature and other sources of reliable information on existing conditions (the affected environment) but rely excessively on speculation when discussing potential impacts. The best EAs rigorously document scientific papers addressing possible effects of actions on affected environmental receptors. For example, an extensive base of peer-reviewed scientific papers is available that discuss potential effects of wastewater discharges and salt drift from power plants on terrestrial and aquatic biota.

As for EISs and other scientific writing, EAs should reference primary sources of scientific information, especially peer-reviewed scientific literature (usually journal articles) whenever possible. Published scientific review papers are also a reliable secondary source of scientific information that can be used when primary literature is not readily accessible or excessively voluminous. Federal and state agencies publish a wealth of technical books, papers, open-file reports, annual status reports, and pamphlets covering almost all major environmental technical issues. From a scientific research perspective, this body of work is sometimes termed the "gray literature." Its technical reliability is highly variable; some items approach the rigor of peer-reviewed papers, some are of good technical quality but perhaps biased toward the interests of the originator, some may contain minor errors or exaggerations, and the worst can be purposefully erroneous or misleading. Researchers should carefully weigh the quality of available information sources and when possible compare technical information in one source against similar information in others. Preparers of EAs (as well as EISs) should, however, remember that NEPA documents are public reports and not scientific theses or scientific research papers; authors should not arrogantly limit their scope of research to the formal scientific literature and overlook other useful information sources.

An especially prevalent secondary information source cited in NEPA documents is previous NEPA documents. Although previous NEPA documents are a logical source of information, preparers of new EAs (and EISs) should carefully consider the reliability of information presented in earlier NEPA documents, which vary widely in quality. Unless the proposed action in the current EA fulfills a relatively complex proposal, affected environment descriptions lifted from previous EISs are generally too long to serve the same purpose in an EA and should be condensed. NEPA preparers should strive to read the references cited in the earlier NEPA documents and cite the primary sources when possible.

The World Wide Web is perhaps the grayest of the gray literature. Publishing Web pages is incredibly easy, almost as easy as sending an e-mail (and for many people today, easier than writing a letter). The bar is very low; no editors, reviewers, page charges, or reproduction costs. This is not to say, however, that good, reliable technical information is not available from Web pages. Web pages associated with scientific professional associations or government agencies are usually of good quality, those associated with advocacy groups may be of good quality but biased, while "personal" Web pages and blogs constitute little more than electronic conversation. Many Web pages contain links to high quality papers; in such cases the researcher should seek and cite the linked paper rather than just reading a Web page summary and citing the URL. Authors relying on information obtainable solely from a Web page must be extremely careful to authenticate the reliability of the information. Because even good quality Web pages are often changed frequently, authors citing Web pages should include the access date in the reference information.

One distinct advantage of Web pages as data sources is that the EA (or EIS) text can contain hyperlinks, thereby providing easy access to referenced sources. Very soon, NEPA documents will primarily be published on the Web. However, in the meantime, preparers must still assume that some readers will read only on paper, and thus not be able to view hyperlinks.

7.2.4 Documenting assumptions

Future uncertainties are often dealt with by making reasonable assumptions. A common criticism of NEPA documents is inadequate referencing of source material, including assumptions. Regardless of length, every EA, just like an EIS, should include a list of all references used in the technical analysis. Although not specifically required for an EA under the CEQ guidelines, agencies should maintain an administrative record of all source documents used in preparing each EA and supporting technical documents. One often overlooked but key reference item consists of calculation sheets used to generate numerical data presented in an EA; all data presented in an EA or EIS need to be traceable to a source. As for an EIS, references that are not easily accessed by the public may need to be included with the EA as appendices. The ability in recent years to present appendices, and sometimes the entire EA, on electronic media rather than paper has made it easier to include reference material with the EA text.

The EA must clearly identify any uncertainties and the assumptions used to bridge them. The rationale on which each assumption is based should be clearly documented. Such practice can be critical in defending the validity of the analysis, if challenged. Formal protocols for conducting uncertainty analysis have been developed for the human health and ecological risk assessments conducted to support decision making for cleanup of environmentally contaminated sites under the Comprehensive Environmental Response, Compensation, and Liability Act (CERCLA; commonly called Superfund). Formal uncertainty analysis has not commonly been applied to NEPA, especially to EAs, but could provide valuable insight and improved understanding of potential environmental impacts. The uncertainty protocols developed for Superfund involve tracing sources of uncertainty encountered at each step of analysis. For example, uncertainty sources in an analysis could include component uncertainties inherent in source data used, modeling protocols followed, and interpretation of modeled output. Uncertainties associated with earlier steps in such a linear analytical process can propagate with each subsequent step, contributing to progressively larger overall uncertainty. Although both qualitative and quantitative uncertainty analyses are performed for Superfund risk assessments, most uncertainty analyses for EAs would likely be qualitative only.

7.2.5 "Will" versus "would"

To clearly demonstrate that no decision has yet been made, discussions of potential actions should be written as if the action might take place. Words such as "proposed," "would," "could," and "might" should be used instead of "will" to clearly indicate that a final decision generally has not been made. In a similar fashion to an EIS, an EA is prepared to support a "determination" and is thus not itself a decision document. For an EIS, the decision is the choice of an alternative, and it is documented in a record of decision (ROD). For an EA, the determination is whether to issue a finding of no significant impact (FONSI), which is a decision to proceed with an action without further NEPA analysis. If a FONSI cannot be issued, the agency must prepare an EIS or modify or abandon the action.

As a result of this distinction, an EA should not be worded as if a FONSI is a foregone conclusion. In fact, the wording in the regulations clearly indicates that some EAs may lead to a determination to prepare a subsequent EIS rather than a FONSI.

7.2.6 Readability and plain language

Write to express—not to impress. NEPA documents, including EAs, are public documents and the public must be able to read and understand them. If citizens cannot understand the material in the assessment, they cannot effectively participate in the NEPA process; this can lead to distrust, resentment, hostility, and even lawsuits. Reader-friendly documents

Table 7.1 Seven simple techniques for improving readability

1. Provide an overview.
2. Provide clear headings.
3. State headings as questions.
4. Make headings distinct.
5. Use locally recognizable landmarks to identify the location(s) of future actions.
6. Explain technical terms as they come up (rather than in a glossary).
7. Use text bullets.

tend to foster public understanding and cooperation, and lessen suspicion. For instance, technical supporting data can be briefly summarized in simple terms in the EA with the technical detail provided in the appendix or incorporated by reference from publicly available material.

To increase readability, practitioners should make good use of graphic aids (maps, tables, figures, graphs, and flowcharts) to enhance a reader's comprehension. Table 7.1 lists some simple techniques for improving readability.

7.2.6.1 Style

There is no required writing style for an EA. The CEQ NEPA regulations do, however, require that the analysis be written in plain language (40 CFR 1502.8), that it be concise, and that it be analytic, not encyclopedic (40 CFR 1502.2). It should also be based on scientific accuracy and reflect known information (40 CFR 1502.24). In addition, a document's organization plays a significant role in the overall quality of the document and its effectiveness in conveying the primary message. The following guidelines on style will assist the NEPA staff in preparing quality NEPA documents:

- Define all abbreviations and acronyms the first time they are used in the document.
- Write EAs precisely and concisely, using plain language. Refer to: http://www.plainlanguage.gov/ for information on plain language.
- Ensure information provided in tables and figures is consistent with information in the text and appendices.
- Provide a list of abbreviations and acronyms with definitions at the beginning of the EA.
- Minimize the use of abbreviations and acronyms to the extent practical. In doing so, use only those acronyms that are referred to frequently in the EA or commonly recognized by the public.

Chapter seven: Writing the environmental assessment

- Define all technical terms that must be used, preferably in a single glossary or definitions chapter.
- If scientific notation is used, provide an explanation.
- Use consistent units of measurement throughout the document.

While CEQ and most agency NEPA regulations mandate no specific style for an EA, individual offices and branches of some agencies have clear preferences on writing style. Preparers, especially contractors, should discuss general writing styles with the individuals who will review the EA. Those individuals may indicate a specific style manual to follow.

7.2.6.2 Plain language

The use of plain language is encouraged by all federal agencies throughout their documentation and is not limited to just NEPA. As part of his well-publicized effort to "reinvent" government, President Clinton issued a memorandum on June 1, 1998, encouraging the use of plain language in all agency communications with the public, including all proposed and final rulemakings published in the Federal Register. More recently, President Obama has further promoted plain language as a means of promoting transparency in the government.

Schmidt has proposed eight plain language questions that he believes will lead to a "perfectly" prepared EA (Table 7.2).*

7.2.7 Visuals and graphic aids

Anyone who has read a newspaper, magazine, brochure, investment prospectus, annual report, or a Web page knows that text is but one part of the overall written communication package. The same is true for EAs and EISs. Carefully designed maps, illustrations, photographs, tables,

Table 7.2 Eight plain language questions to ask when preparing an EA

1. What are they up to?
2. Why are they doing that?
3. What else would do the same thing?
4. What's so bad about doing nothing?
5. What are the comparative merits of each alternative?
6. On what basis will a decision be made?
7. What, if anything, will be done about the adverse consequences?
8. What monitoring will be done, if any?

---

* DOE, NEPA Lessons Learned, "Preparing Focused and Concise EAs," June 1, 2009; Issue No. 59, pp 22, Second Quarter FY 2009.

and other visual aids can convey volumes of information that might have been impossible using text alone or that would have required voluminous descriptions. Each should be called out and integrated with the text. The accompanying text should discuss and interpret but not repeat the information presented in the visual aids.

How graphic aids increase comprehension was the topic of two published articles in *Environmental Impact Assessment Review* discussing how graphic aids can be used to increase comprehension.[1] Three researchers from the University of Illinois conducted tests on students to measure their ability in understanding and recalling the project description and environmental consequences of a local flood-control plan in an EIS. In the first study, students read portions of the EIS and then answered questions about the project and its environmental effects.[2] The study concluded that the students' understanding of the material was "atrocious," even among the best readers. The students' performance was generally below 70 percent—a measure the authors of the study considered adequate for comprehension (the equivalent of an academic "C").

Pictures of a local creek were used to show how the creek would look if flood control measures were implemented (see Figure 7.1). Comprehension of the proposal and the understanding of environmental impacts improved to a level significantly greater than 70 percent.

7.2.7.1 Figures

Advances in geographic information system (GIS) technology has greatly facilitated development of effective graphics (commonly termed "figures") for NEPA documents. Brief, low-budget EAs that might have once been prepared without graphics or with only simple graphics can now include visually effective maps and figures. The distinction between graphics development and technical analysis has been greatly blurred. The GIS layers that are now developed as the basis for area calculations once performed manually using disk integrators and other time-consuming antiquated equipment can rapidly be developed into easy-to-read graphics.

Additionally, while some agencies still insist on black and white graphics in NEPA documents, the use of color greatly enhances readability of graphics. While color printing remains expensive, publication of EAs and other NEPA documents on websites and electronic media allows for inexpensive dissemination of color images. While not yet common, future EAs and other NEPA documents will probably make more extensive use of embedded color photographs to enhance information conveyed in text.

7.2.7.2 Tables

Using tables to present numerical data or large volumes of text data can be another technique for enhancing data presentation and condensing

Chapter seven: Writing the environmental assessment   145

Figure 7.1 Potential flood control features suggested for Hickory Creek involved three different changes of the creek banks. The banks were to be changed from their existing condition (top left) to either a fabric-formed concrete embankment (top right), a vertical concrete wall (lower left), or an earthen embankment (lower right). Photos from Sullivan, W.C., Kuo, F.E., and Prabhu, M. Communicating with citizens: The power of photosimulations and simple editing. *Environmental Impact Assessment Review*, 17(4): 295–310, July 1997. Reprinted by permission of Elsevier Science, New York.[4]

text. Use of tables in NEPA documents has increased in recent years with the increased availability of spreadsheet and database programs. But excessive use of tables, especially tables of numbers, can overwhelm readers. Huge multipage tables presenting concentrations of dozens of chemical constituents in environmental media samples may be appropriate in technical environmental reports such as remedial investigations under Superfund. But those tables should be condensed if possible to highlight only relevant data or trends when used to address chemical contamination or waste management issues in an EA or EIS. It is sometimes possible to summarize key points from a large table in one or two sentences of text. The text can then direct the reader to the source report to see the actual table. If large or numerous tables are not readily accessible from primary sources and must be included in the NEPA document, they can be placed into an appendix rather than inserted in the text.

7.2.7.3   *Photographs and photosimulations*
The availability of software capable of pasting photographs into text and the reduction in color copying costs have made the use of color

photographs easier in EAs (and EISs and other environmental documents). The same is true of software capable of projecting scaled images of proposed structures onto photographs of the existing landscape (photosimulation). Software is available that uses optic modeling calculations to precisely calculate the apparent size of a structure to a viewer standing at a specific vantage point of interest (e.g., a house, park, or historic place of interest) at a specific distance and direction. It is not an artistic process of "winging it," although it employs many of the optical and physical processes known for centuries to painters and sketchers aiming for realism.

Photosimulation is an especially valuable tool for demonstrating the aesthetic visual effects of proposed tall structures such as tall buildings, smokestacks, communications towers, or electric transmission line towers. Embedded photographs and photosimulations can greatly enhance the readability of NEPA text, although many agencies still prefer to segregate photographs into appendices. As with other visual aids, photographs should complement and expand upon information in the text, and extraneous photographs should be avoided. Extensive photographic logs of site visits can be maintained in project files or the administrative record rather than included in an EA (or EIS).

7.2.7.4 Other techniques for improving readability

The Illinois researchers' second suggestion for improving comprehension involves better editing. Practitioners can improve the effectiveness of their documents by following seven simple techniques: provide an overview, provide clear headings, state some headings as questions, make headings distinct, use locally recognizable landmarks to identify locations of project work, explain technical terms as they come up (in addition to a glossary), and use text bullets. When combined with photosimulation, these seven simple techniques increased students' comprehension to more than 80 percent.

7.2.8 Glossaries

An easy and effective way to make NEPA reader-friendly is to mark in bold or italics the first occurrence of terms that are defined in the glossary. This will effectively signal the reader to consult the glossary. This system should be explained in a footnote or text box at the beginning of the NEPA document and also in the glossary. Glossaries are common features in EISs but rare in EAs. In general, the need for a glossary is greatest for longer and more complex EAs. Technical terms can usually be defined directly in the text in shorter EAs.

7.2.9 Other writing style notes

Two additional items related to writing style and effective written communication in EAs and other NEPA documents deserve special mention: use of measurement units and use of Latin scientific names for species of plants and animals.

7.2.9.1 Measurements

The first concerns units of measurements. Most scientists in the United States prefer using metric measurements, while the general public in the United States is most familiar with U.S. customary measurements. Many EAs and other NEPA documents use one type of unit with the other units in parenthesis, i.e., "The road would be 1.6 kilometers (mile) long" or "Approximately 2.5 acres (1.1 hectares) of land would be disturbed." This approach makes the text more difficult to read. An alternative approach is to present each measurement using only one unit and include a conversion table as an appendix. The choice of units should consider what is understandable to the widest possible audience (this is not always the objective of scientific journal articles, which mostly require use of metric units). U.S. customary units are usually most recognizable to the public for basic parameters such as length (miles), area (acres), and volume (gallons). However, metric units are most recognizable for laboratory measurements such as milligrams per kilogram (mg/kg).

7.2.9.2 Species names

The second item is the use of scientific names (sometimes referred to as Latin name or binomial) to identify a plant or animal species. Biologists commonly refer to plants and animals using two-word scientific names comprised of a genus followed by the name of the species. For example, the scientific name for the tree species red maple is *Acer rubrum*. *Acer* is the genus, which includes all of the trees and shrubs commonly referred to as "maples." *Rubrum* is the specific epithet for red maple (notice that it is the Latin for "red"). Some people use other common names such as swamp maple for red maple, but *Acer rubrum* unquestionably refers to only one species. Most biological technical reports refer to red maple as "red maple (*Acer rubrum*)" at the first mention of this species but simply as "red maple" thereafter. A few technical reports use scientific names exclusively, although even most peer-reviewed scientific journals follow the first approach.

EAs and other NEPA documents should always use common names recognizable to the general public. Some NEPA practitioners feel that scientific names should not be used in NEPA documents. But because the documents have to be readable to a national audience, inclusion of scientific names following the first mention of a species can help add clarity and avoid confusion inherent in regional variation of common names.

Alternatively, an EA can include an appendix listing the scientific name of each plant or animal mentioned in the text. Note that scientific names are always italicized, as they are words from a foreign language (Latin).

7.2.10 How long should the EA be?

The CEQ views an EA as a "concise public document." It should not contain long descriptions or overly detailed data. Rather, CEQ states that it is to be a concise document that provides a brief discussion of the need for the proposal, alternatives, and environmental impacts (§1508.9).

While the regulations do not contain page limits for an EA, the CEQ has advised agencies to keep the length of an EA to a maximum of 10 to 15 pages.[5] As explained previously, the length of an EA is less important than the content of the EA: An EA should be just long enough to present the relevant information allowing decision makers to evaluate environmental impacts from proposed actions and reasonable alternatives and determine if a FONSI is justified. To avoid undue length, the EA may incorporate by reference background data to support its concise discussion of the proposal and relevant issues.

7.2.11 Editing

Most agencies use a technical editor to review draft and final EISs prior to publication. The editor is usually an agency employee, employee of a contractor preparing the EIS, or a freelance specialist hired by the agency or subcontracted by the contractor. A good technical editor for an EIS combines broad but generalized technical environmental knowledge with a keen understanding of how to use language and visual aids to convey information to nontechnical readers. Most EAs are shorter documents than EISs and are often prepared without the involvement of a technical editor. However, technical editing is a valuable contribution to EAs. In fact, a good technical editor can help to further condense the text of an EA to achieve the desired brevity without compromising content. Perhaps the most beneficial, and most difficult, job of a technical editor in NEPA is to ensure consistency of technical information and data throughout documents that tend to be segmented into disparate technical issues addressed by different specialists.

7.3 Typical contents of an environmental assessment

As indicated in Table 7.3, the CEQ regulations specify only four documentation requirements that an EA must satisfy (§1508.9[b]). The first three

Table 7.3 CEQ documentation requirements for an EA (§1508.9[b])

1. Need for the proposal.
2. Alternatives (as required by Section 102[2][E] of NEPA).
3. Environmental impacts of the proposed action and alternatives.
4. Listing of agencies and people consulted.

items indicated in Table 7.2 are common to both the EA and EIS (although the breadth and level of detail may not be as great for most EAs). The fourth item is required only in EAs (but can be useful in an EIS). The regulations do not specify a particular outline for the EA, and they do not limit what an EA may include.

Each agency's internal guidance and NEPA implementation procedures should be consulted for any additional documentation requirements.

7.3.1 Suggested EA outline

To improve decision making and promote a more open public process, it may be advantageous to include information beyond CEQ's minimal requirements. However, it is important to strike a balance between the benefits of additional information and the expense resulting from increased work scope and production costs. Contractors tasked to prepare EAs (or bidding to prepare EAs) should not assume that agencies will do more than the minimum required. When preparing RFPs for NEPA services, agencies should carefully indicate any items that exceed minimum requirements as set by CEQ and the agency's own regulations.

A generalized outline for balancing these competing objectives is presented in Table 7.4. This outline meets the minimum requirements of §1508.9[b] while including other elements contributing to effective communication without necessitating substantial additional effort. It reflects general organizational elements in common use by many agencies when preparing long or complex EAs (EAs for very simple projects may follow shorter templates). The outline for any EA should be tailored to meet the agency's specific mission and circumstances, including (as appropriate) any specific requirements cited in the agency's orders and NEPA implementation procedures.

Including a title page, glossary, and executive summary is consistent with the Seven Methods for Increasing Readability (Table 7.1), while adding only marginally to the document size.

The reader should note that an EIS must state both a purpose and need for taking action (see first item in Table 7.2). The CEQ specifically calls for need but does not mention purpose for an EA. Discussing the purpose in addition to the need in an EA may help explain the project's rationale to the public and decision maker.[7]

Table 7.4 Suggested outline for an environmental assessment

1. Title page
2. Glossary
3. Executive summary
4. Purpose and need for the proposed action
5. Description of the affected environment
6. Description of the proposed action and reasonable alternatives
7. Analysis of environmental impacts of the proposed action and reasonable alternatives
8. Applicable environmental permits and regulatory requirements that would need to be obtained
9. List of agencies and people consulted
10. List of preparers
11. References

Most EAs are not long enough to warrant an index. Nevertheless, unusually long EAs might benefit from an index (although they might benefit more by an effort to condense their text).

7.3.2 Organization

While all are familiar with CEQ and agency regulatory requirements regarding the content and organization of EAs, agencies, offices within agencies, contractors, and even individual NEPA practitioners on agency and contractor staffs have varying and distinct approaches toward organization of an EA. The general organization of EAs generally differs more than that of EISs, for which CEQ has issued more detailed guidance. The authors recognize the following competing concepts for approaching the writing of affected environment and environmental consequences in EAs:

- Addressing the affected environment and environmental consequences requirements in a combined chapter.
- Addressing the affected environment and environmental consequences requirements in separate chapters (traditional format).
- Discussing environmental effects on an alternative-by-alternative basis.
- Discussing environmental effects on an affected resource-by-affected resource basis.

Any of the aforementioned general approaches (and combinations thereof) are acceptable, but their effectiveness and efficiency are highly dependent on the complexity of the action being taken. The EA preparers should carefully consider which of these presentations is most appropriate for a particular EA. Some guidelines to consider regarding organization of EAs include:

- Be consistent in how the effects on environmental resources are analyzed (choose one organizational scheme).
- Describe the net environmental effects, or residual impacts, in summary form at the beginning or end of the discussion.
- Summarize net effects in tabular form to allow ease of comparison across alternatives.
- Present alternatives and resources in the same order throughout the document.
- Present the no-action alternative first to establish a baseline against which other alternatives will be compared.

7.4 Performing the analysis

Consensus must be reached regarding the appropriate use of any analytical methodologies or computer models that are employed. Where possible, analysts should use methods and computer models endorsed by the U.S. Environmental Protection Agency (EPA), other governmental agencies, or professional societies.

7.4.1 Five-step methodology

"Practitioners are mandated to "utilize a systematic, interdisciplinary approach ... in planning and in decision making which may have an impact on man's environment."[8] To this end, a five-step methodology for investigating environmental impacts is offered in Table 7.5. Professional discretion must be exercised in determining how these steps are most practically applied based on the complexity of the analysis and the particular circumstances. The reader is referred to Section 2.4 of the companion text, *NEPA and Environmental Planning*, for additional information.

Step 1: Characterize potentially affected resources—The term "potentially affected environmental resources" refers to the existing physical, biological, cultural, and socioeconomic resources before they are altered by the agency's proposal. These resources should be identified and characterized prior to beginning the environmental impact analysis. The characterization should not, however, be exhaustive; it should only provide the minimum baseline information needed to support the analysis. More than any other part of an EA (or EIS), the section describing baseline conditions is the most subject to excessive "encyclopedic" detail resulting in excessive length.

Commonly consulted sources of information. The affected environment descriptions in many EAs utilize technical information from standardized

Table 7.5 Systematic methodology for performing an EA analysis

Step 1—*Characterize Potentially Affected Resources:* Data is collected on the potentially affected resources. Temporal and spatial boundaries are delineated for the subsequent analysis. The potentially affected environmental resources are described.

Step 2—*Identify Component Actions*: All component actions that may occur over the reasonable foreseeable life cycle of the proposal are identified.

Step 3—*Identify Environmental Disturbances:* Environmental disturbances (i.e., noise, effluents, emissions, ground disturbances) resulting from the component actions are described. Cause-and-effect relationships are described.

Step 4—*Screen and Analyze Environmental Impacts:* The environmental disturbances are investigated to determine how they would *impact* (change or affect) environmental resources. The environmental disturbances are screened to determine their potential for significance. Disturbances deemed to be clearly and unequivocally nonsignificant are eliminated from further study. A more detailed analysis is then performed to determine how the remaining environmental impacts might significantly impact environmental resources.

Step 5—*Assess Potential Monitoring and Mitigation Measures:* If appropriate, potential mitigation measures are evaluated to determine their effectiveness in mitigating potentially significant impacts. A monitoring plan can also be prepared.

sources available for most locales throughout the United States. EAs generally rely on easily accessible information sources to a greater extent than EISs, which often require substantial site-specific research and information from tailored site investigations. Most of these sources are published by federal, state, and municipal agencies. Many are now available online. Otherwise, the data can often be rapidly obtained from the publishing agency or from local offices at no charge or for a nominal charge. Other sources can be requested on the telephone or in writing. Analysts often visit local government offices to obtain information sources while on a trip to visit the proposed site.

Table 7.6 lists some of the technical sources that are routinely consulted in preparing EAs. The list is not exhaustive. Readers are urged to consider other sources of technical information that might also be available and appropriate for the specific action in question.

Many of the databases in Table 7.5 are now available as online GIS layers where users can interactively outline polygons and quickly generate site-specific data, usually at no charge. Examples include soil survey layers, floodplain layers, and wetlands layers. EA preparers using online GIS layers should carefully document how they defined their polygon(s) and the date of access to the GIS layer.

A variety of site-specific environmental documents often provide useful technical information for analyzing actions taking place within

Table 7.6 Routinely consulted technical information sources

Source	Available from	Relevant to
Soil surveys	Natural resources Conservation Service	Soils geology
National wetland inventory maps	U.S. Fish & Wildlife Service; often available from municipal offices	Wetlands, biological, and water resources
Flood insurance rate maps	Federal Emergency Management Agency; usually available from municipal offices	Floodplains and water resources
Topographic maps (7.5 minute)	U.S. Geological Survey	Topography water resources
Topographic maps (1- or 2-foot contours)	These more detailed topographic maps are sometimes available from municipal offices	Topography water resources
Comprehensive plan (municipal or county)	Municipal offices	Land use; socioeconomics; biological resources
Zoning maps	Municipal offices	Land use
Aerial photographs	Municipal offices; National Aerial Photography Program	Land use; biological resources
STORET (storage & retrieval) water quality database	U.S. Environmental Protection Agency (can access free online)	Water resources
Water-data reports	U.S. Geological Survey	Water resources
Environmental record database searches	Sold at nominal cost from private-sector database search companies	Hazardous materials/ waste management
Census data (population, employment, etc.)	Bureau of the Census (can access free online)	Socioeconomics
State natural heritage databases	State natural resource management agency; must access in writing and might include a nominal charge	Threatened and endangered species; biological resources
State Historic Preservation Officer	State cultural resource agency; access in writing as part of Section 106 consultation process	Cultural resources

existing federal installation boundaries. These may include previous EAs and EISs, site environmental reports, natural resource management plans, environmental baseline surveys, and geographical information system layers. Readers are also advised that at least one site visit is recommended regardless of how much published technical data is available.

Several commercial companies now offer environmental data from computerized record sources. These companies were primarily formed to provide rapid database searches to facilitate preparation of Phase I environmental site assessments addressing the potential for environmental contamination of property. Recently, however, they have expanded their database capabilities to offer information useful in preparing NEPA analyses. Sources of data include:

- Data on sites listed in the National Register of Historic Places.
- Wetlands mapped on National Wetland Inventory maps.
- Floodplains mapped by the Federal Emergency Management Agency.
- Threatened and endangered species.

The services offered by these companies are typically very fast (on the order of one to two days) and relatively inexpensive (typically less than $500).

The reader is cautioned that some of the information obtained from these database searches may be incomplete or inadequate. For example, sometimes only information on federally recognized historic and archaeological sites or on federally listed threatened or endangered species is available. Information from state or local lists is often not provided.

Furthermore, the database searches may provide inadequate descriptive detail concerning the features noted. For example, a review of National Wetland Inventory maps does not constitute an adequate determination of whether wetlands exist on a given site. Perhaps most important, merely reviewing computerized database information fails to engage environmental agency personnel in the early planning process required in preparing an NEPA analysis.

While EA preparers should not be discouraged from using these environmental database search services, they should understand the limitations of the searches. The searches do not replace the need for an independent and thorough investigation by analysts preparing the NEPA analyses. Moreover, such searches do not meet the agency consultation requirements under NEPA.

Practitioners are cautioned to use the searches as a supplemental research tool; the practitioner is responsible for cross-checking the accuracy and supplementing the data as necessary to adequately support the impact assessment. As appropriate, regulatory agencies must still be contacted to determine whether additional relevant information is available.

NEPAssist. The EPA maintains a nationwide GIS application called NEPAssist, which displays several sets of environmental data spatially. The GIS allows analysts to perform a "virtual site visit" of a proposed project. Potential benefits include: identifying important environmental issues at early stages of project development, focusing on significant environmental impacts, helping direct project siting to areas that are the least environmentally sensitive, and facilitating collaboration during the preparation of NEPA documents. Using data layers of the project site map, users can outline the footprint of a proposed project on a map, and then use the NEPAssist application to generate a report identifying nearby resources (e.g., number of streams and wetlands, 100-year floodplain?).

Practitioners have reported that NEPAssist has allowed them to identify proposed sites and alternative sites for a proposed action. A public comment concerning disproportionate impacts on a minority community adjacent to a proposed site led one analyst to choose the Environmental Justice Demographic Mapping Tool data available through NEPAssist. This allowed the user to assess demographic, health, economic, and employment data (and compare site data to that for the county and state) at various distances from alternative sites to more effectively understand potential concerns.

In another case, scoping concerns related to proximity of wetlands and potential flooding led the user to access U.S. Fish and Wildlife Service wetlands and Federal Emergency Management Agency floodplain data layers. This allowed the user to swiftly determine whether the project would be located in a wetland or floodplain, and whether a wetland and floodplain assessment would be needed in the EIS. For direct inquiries or to apply for a password to NEPAssist, contact Aimee Hessert, EPA Office of Federal Activities, at hessert.aimee@epa.gov or 202-564-0993.

Step 2: Identify component actions—NEPA recognizes three types of action: (1) connected, (2) cumulative, and (3) similar.[9] A final course of action (identified as either the "proposed action" or one of the "alternative actions") will invariably consist of a set of individual or component actions. Each component action will be related to other component actions by either being connected, cumulative, or similar. To adequately evaluate environmental impacts, all reasonably foreseeable component actions must be identified and described. For example, a proposal involving construction of a small office building might involve an array of component actions such as site clearing and excavation, construction of access roads, construction of power and communications cables, construction of a water and sanitary line, a parking lot, and drilling a water supply well.

Agencies must strive to avoid knowingly segmenting component actions, preparing separate NEPA documents for each. Segmentation is an especially nettlesome problem in EAs, where segmented components may lack potentially significant environmental impacts and qualify for

FONSIs but where the totality of impacts from the complete action exceeds the bar for significance. Common sense and logic must be applied; separate activities conducted in response to distinctly different needs should definitely be treated in separate NEPA documents, while splitting a road or other linear project into small but generally similar segments solely for the purposes of keeping impact levels in a progressive series of highly similar EAs quantitatively low is clearly incompatible with the objectives of NEPA.

Step 3: Identify environmental disturbances—The terms "environmental disturbances" and "environmental impacts" should not be confused. An environmental disturbance can be viewed as some type of stressor produced by an action that might change or affect an environmental resource. For example, a new aircraft might release high-altitude carbon dioxide emissions (i.e., environmental disturbance). Such a release in itself is not an environmental impact. The environmental impact denotes how such a release might affect or change an environmental resource (e.g., contribute to global warming). Thus an environmental impact depicts how a change in an environmental resource, caused by an environmental disturbance, would affect humans and environmental quality.

Each potential action may produce environmental disturbances. The disturbances must be identified and described in detail sufficient to allow analysts to investigate how they could affect or change environmental resources. Pathways and cause-and-effect relationships should also be identified. The Table 7.7 provides some examples of environmental disturbances and associated environmental impacts.

The boundary between environmental disturbances and environmental impacts is not always sharply defined. For example, the clearing of 10 acres of natural habitat constitutes both an environmental disturbance (in that the clearing is a part of the project description) and an environmental impact (the loss of habitat for plants and wildlife). Distinguishing environmental disturbances and impacts is not essential. But ensuring that an analysis proceeds beyond description of disturbances to identification of resulting impacts is essential. Most EAs (and EISs) for construction proposals quantify wildlife habitat losses; only the best EAs follow through to identify how the habitat losses might affect wildlife populations and behavior in the landscape.

Step 4: Screen and analyze environmental impacts—The environmental disturbances are screened (e.g., scoping procedure) to determine how they would affect environmental resources. The potential disturbances can be compared with quantitative and qualitative metrics (significance criteria) to determine whether they are clearly inconsequential. Disturbances deemed to be clearly and unequivocally nonsignificant are eliminated from more detailed study. A more detailed analysis is then performed

Table 7.7 Examples of environmental disturbances and associated environmental impacts

Environmental disturbance	Related environmental impacts
Clearing natural vegetation from 10 acres of land	Loss of 6 acres of deciduous forest habitat and 4 acres of old field habitat
	Severing of a forested corridor that presently facilitates movement of forest-dependent wildlife
	Eliminating known nest site for bald eagles
Generating brief and irregular periods of construction noise exceeding 90 dBA	Startling wildlife, causing abandonment of nests and habitat
	Inhibiting enjoyment of natural surroundings in adjacent parklands
Employing 2,000 skilled construction workers over a 2-year construction period	Creating a short-term shortage of hotel accommodations
	Creating Level F traffic conditions during specific morning and afternoon hours on a stretch of roadway approaching the site

to determine how the remaining environmental impacts would affect environmental resources.

A no-action alternative provides an effective environmental baseline for performing the analysis. Thus analysts have a basis for comparing the effects of pursuing a particular action against the impacts of taking no action. Metrics can be used to determine whether a certain level/degree of change from the baseline level is deemed significant. Once environmental impacts of taking no action are determined, an effort is mounted to determine the impacts of the proposed action (and other reasonable "action" alternatives).

Step 5: Assess potential monitoring and mitigation measures—If deemed appropriate, potential mitigation measures are also evaluated. The reader should note that a nonsignificant impact does not necessarily imply that there would be no impact—only that the impact is below the threshold of significance. A monitoring plan may also need to be prepared.

7.5 Specific documentation requirements and guidance

Some aspects of a proposal might be considered sufficiently covered by merely mentioning that the specific action will take place. In others cases, an extensive analysis may be necessary. Since NEPA's inception, no definitive direction has been established for determining the amount of detail, discussion, and analysis that is sufficient to adequately cover a proposed action. Yet, agency decision makers are routinely called upon to do just that. Inevitably, such determinations tend to be subjective. A tool (Sufficiency-Test) is offered in Chapter 2 of the companion text, *NEPA and Environmental Planning*, for assisting practitioners in dealing with this issue.[10]

Specific documentation requirements and guidance governing preparation of an EA are provided in the following sections.

7.5.1 Need section

As indicated in Table 7.8, the regulations require that the EA indicate the need for taking action. Interestingly, the regulations do not require that an EA indicate the purpose for taking action, as is required for an EIS. However, the purpose and need for an action are closely interrelated, and the process for determining the need for a proposed action addressed in an EA is basically the same as the process for determining the purpose and need for an action addressed in an EIS.

In the past, a recurring problem has involved discussing the need for preparing the EA rather than the need for taking action (i.e., proposal). For example, some EAs have mistakenly explained that the need was to comply with NEPA. Underlying is the inherent need for the benefits and services provided by the action; the "need" to comply with NEPA is merely a consequence of having decided that there is a justifiable basis for proceeding with the action.

A simple technique that practitioners can use in determining the underlying need for a proposed action involves asking the following question: Why is the agency considering the proposed action?

"Need," as defined in Webster's Dictionary, is "a lack of something useful, required, or desired."[11] Put succinctly, a need is something lacking

Table 7.8 Simple examples of possible needs underlying some common types of actions addressed in EAs

Action	Underlying need
Widen a segment of a federal highway	Improve traffic flow and allow more heavy trucks to access an existing industrial park
Remove trees at the end of a runway	Improve air traffic safety by providing greater clearance in the approach and departure airspace
Construct a power plant	Meet projected baseload power requirements for a defined region

or desired. As espoused by Schmidt, the statement of the purpose and need is critical in successfully identifying the range of reasonable alternatives for later analysis.[7,12] Properly defined, the statement of need provides an effective tool for screening out what might otherwise be a diverse or even unbounded range of alternatives.

A broadly written statement of need can facilitate agency planning, as it tends to compel consideration of more diverse approaches for satisfying the agency's need. The disadvantage is that it can increase the scope of analysis, because a greater number of alternatives might need to be investigated. If the number of alternatives is unwieldy, this range can be reduced by narrowing the statement of need. However, the reader is cautioned that it may be inappropriate (and possibly illegal) to define the statement of need so narrowly that it essentially eliminates all reasonable alternatives but the proposed action.

7.5.2 Affected environment section

As indicated earlier, the regulations do not require a description of the affected environment in an EA. However, inclusion of an affected environment section can be very useful, as it provides a baseline description of the existing environment, against which both the intensity and context of potential impacts can be compared. The affected environment discussion should briefly describe the condition of the affected environmental resources prior to implementation of any alternative.

Lacking such a baseline, it may be difficult or impossible to conclusively demonstrate that the potential impacts are truly nonsignificant. An EA, like an EIS, must succinctly describe the environmental resources potentially subject to impact.

The EA should succinctly describe potentially affected environmental resources. Each resource should be described in only enough detail to allow the reader to understand how it could be affected. Consistent with a sliding-scale approach, such discussions should be commensurate with the potential for sustaining a significant impact. As warranted, the discussion and data may be incorporated by reference (§1502.15).

Introduction to the affected environment section—The introduction to the affected environment section should indicate the purpose for which this section has been prepared (i.e., to provide a baseline for gauging impacts). The location or area that would be affected should be clearly delineated on a map. The discussion should be presented on a resource-by-resource basis, preferably in the same order as resources are evaluated in the section describing the environmental impacts of the proposal.

Interpreting baseline measurements—Where possible, indicators should be provided that can be used in gauging the environmental impacts that will be described later. For example, Table 7.9 indicates hypothetical

Table 7.9 Hypothetical background noise levels at four alternative locations

Location	L_{eq}-24 (dBA)
Proposed location	39
Alternative location No. 1	45
Alternative location No. 2	34
Alternative location No. 3	63

baseline noise levels (measurements are in equivalent sound level [L_{eq}]) that have been measured at four alternative siting locations. To assist the reader in interpreting these measurements, Table 7.10 indicates typical noise levels (measured in dBA) encountered in daily life and industry.

Defining a region of influence—Some NEPA preparers consider it helpful to formally define a region of influence (ROI) that spatially outlines an area within which impact analysis will be performed for a resource. Such an approach introduces discipline to the analysis, and formally describes the context within which the impacts are to be considered. However, such an approach can cause analysts to focus too broadly on impacts rather than emphasizing the local context as directed by the regulations. The rigor of defining an ROI is more common to EISs than EAs, and more common for certain resources (e.g., socioeconomics) than others. But if an ROI is defined, the affected environment discussion should encompass conditions for the resource throughout the ROI.

Needs for affected environment brevity in EAs—The CEQ recognized the tendency toward excessively lengthy affected environment discussions in the 1970s. The affected environment discussion is the part of an EA or EIS most prone to excess length. Affected environment descriptions

Table 7.10 Typical decibel (dBA) values

Sound	dBA
Rustling leaves	20
Room in a quiet dwelling at midnight	32
Conversational speech	60
Busy restaurant	65
Vacuum cleaner in private residence (at 3.1 meters)	69
Beginning of hearing damage for prolonged exposure	85
Heavy city traffic; also a heavy diesel-propelled vehicle (at approximately 7.6 meters)	92
Home lawn mower	98
Air hammer	107
Jet airliner (153 meters overhead)	115

for some resources in many EISs sometimes read like textbooks. Lengthy textbook-like explanations are rarely needed in EISs and are even less essential for EAs.

The need for succinctness is even greater in an EA than in an EIS. The need for any affected environmental discussion at all is limited to resources for which there is substantial uncertainty regarding a potential for significant impacts. Descriptions of existing socioeconomic conditions, regardless of how concise, are usually not needed for EAs not addressing changes in employment. Descriptions of existing terrestrial ecology conditions, regardless of how concise, are usually not needed for EAs not addressing disturbance of naturally vegetated areas.

Although tempting as a cost-saving practice, pasting-in prepackaged "canned" affected environment text (usually from older EAs or EISs) can easily lead to excessive length. The affected environment text for most resource types in most EAs does not need to exceed a few sentences. Writing those sentences from scratch may take less time than pasting and editing previously written text, without the likelihood for including unnecessary details. Furthermore, authors need to be careful to ensure that the information in these older text snippets has not become obsolete.

The "sufficiency test" developed by Eccleston for evaluating the utility of information when writing EISs or EAs is particularly useful when writing affected environment sections of an EA. The test calls for evaluating whether information considered for inclusion in an EIS or EA contributes materially to the ultimate decision making. If the test reveals no contribution, the information can be excluded without compromising the objectives of the document. Although the test appears in that book to be a mechanical process, it really should be an ongoing way of thinking. EA writers should be continually evaluating in their minds whether the material they are writing is truly needed to support their analyses. NEPA documents should not read like textbooks but instead must focus on providing just enough information to support decisions.

A useful rule of thumb is if a piece of affected environment information is needed to assess impacts or compare alternatives, then add it. Otherwise, leave it out.

7.5.3 Proposed action and alternatives section

Consideration of alternatives, including the proposed action, no-action alternative, and range of reasonable alternatives, is introduced in Section 6.3 of this book. The following section focuses on describing alternatives and their impacts in the text of an EA but by necessity overlaps with elements of Section 6.3. Although the CEQ does not refer to this section as the "heart" of an EA as it does for an EIS, a brief but thoroughly thought out and well-communicated section describing the proposed action and

Table 7.11 General-purpose outline for describing alternatives

2.0	Description of Proposed Action and Alternatives	
2.1	Brief Introduction	
	2.1.1	Briefly explain process used to identify alternatives
2.2	No-Action Alternative	
2.3	Proposed Action	
	2.3.1	Location
	2.3.2	Cost and schedule
	2.3.3	Construction activities
	2.3.4	Operational activities
	2.3.5	Support facilities and activities
	2.3.6	Routine maintenance and upgrades
2.4	Alternative A	
	(Similar to that shown for Section 2.3, although alternatives in an EA are usually covered in less detail)	
2.5	Alternative B	

alternatives is invaluable to an effective EA. Section 6.3 discusses in detail the process for identifying alternatives for inclusion in an EA.

A generalized outline for describing the section on the proposed action and reasonable alternatives is offered in Table 7.11 (an expansion of Section 2.0 shown in Table 7.2). As necessary, this generic outline should be tailored to meet the agency's specific mission, circumstances, and requirements.

Proposed action: The necessity for including a description of the proposed action in an EA is obvious. The description must be long and detailed enough to support an environmental impact analysis, but excessive detail detracts from the readability of the document. The description should be factual and unbiased; an NEPA document is not an infomercial or advocacy document. Neither is an NEPA document a design or engineering document. It need not describe the complete design history of a project nor need it embellish on design features not germane to environmental impact assessment. One or two paragraphs may be all that is needed to describe a simple proposed action for a short EA. Even a page or two should be adequate for the most complex project addressed by an EA; a longer description might suggest that an EIS is needed.

No-action alternative: While a description of the no-action alternative is not specifically required by the regulations for an EA, its inclusion is considered good practice, and it might be required by the courts (as a reasonable alternative). The no-action alternative is sometimes erroneously interpreted to mean that nothing will take place in the future that could impact environmental resources in the area under investigation—that the condition of the resources will remain as described in the affected environment text, i.e., remain at the baseline condition. This is not necessarily true: Neither the proposed action nor any related action alternatives will take place, but future activities independent of the subject actions will still occur. A good way of envisioning the no-action alternative is to think of everything considered as part of the cumulative impacts analysis as still occurring, without any incremental impacts from the proposed action or action alternatives. Time does not stand still under the no-action alternative; the environment does not remain static. Things simply proceed without the proposed action or any action alternatives.

Many EAs (and even EISs) include brief but overly simplistic sentences stating that resources would not be affected by the no-action alternative. These simple statements are not always completely true. As a simple hypothetical example, consider an action to build a landfill on a 100-acre tract of old field grassland in the Eastern United States. An EA for that action would have to consider that many bird species favoring such habitats have experienced substantial population declines, and that the proposed action might further contribute to the population declines. But even if the proposed action is not implemented, and the grassland is left intact, it will likely transform naturally to scrub and forest cover through a process termed by ecologists as natural succession. Hence, the habitat losses and associated bird population declines may still occur even under the no-action alternative, albeit more slowly than if ground were suddenly altered by earthmoving and building and operating a landfill there. This impact from the no-action alternatives should be described.

The U.S. Air Force recently completed an EA for removing and trimming trees that had become flight obstructions at the ends of the runways on Andrews Air Force Base in Maryland. The no-action alternative was to not implement measures to address the flight obstructions. However, the Air Force had already implemented a limited program, documented in an earlier EA, to trim certain trees that interfered with use of instrument landing systems used by pilots to land planes on the runways during periods of limited visibility. The tree trimming under the ongoing limited program was less extensive than contemplated under the proposed action, but it was not nil. Nor would it have ceased had the Air Force decided not to implement the more aggressive program needed to treat flight obstructions. Hence, the EA had to indicate that some tree trimming work would take place even under the no-action alternative. The EA had to inform readers that even if the no-action alternative had been chosen, they might still see tree trimmers working on the trees near the runways. In no case would the trees remain at baseline conditions.

Other reasonable alternatives: As for the proposed action, the description of alternatives in an EA should be as brief as possible, including only enough detail needed to support the environmental impact assessment.

Table 7.12 Comparison of plant alternatives

Succinct attributes	Construct new plant	Renovation alternative	No-action alternative
Steam production	Yes	Yes	No
Electric production	Yes (50 megawatts)	No	No
Fuel type	Gas	Coal	Coal
Construction of fuel pipeline	Yes (.6 miles)	No	No
Highway fuel transport	No	Yes (55 miles)	Yes (55 miles)
New plant site	Yes (5 acres)	No	No
Ash sludge disposal	No	Yes (.5 tons/day)	Yes (.8 tons/day)
Requires new construction	Yes, a new plant	Yes, refurbishing an existing plant	No
Water use	50% reduction	No change	No change
Closure of existing plant	Yes	No	No
Land disturbance	Yes, grading	No	No
Demolition of existing structures	25 acres of habitat	Yes	No

The CEQ and most agency NEPA implementing regulations place less emphasis on evaluation of alternatives in EAs than in EISs; hence in an EA, descriptions of alternatives may be briefer than descriptions of alternatives in an EIS. A good approach is to make the length of description for an alternative roughly proportional to how closely the alternative might meet the need established for the proposed action without causing greatly increased environmental impact—in other words how difficult it is to proceed with the proposed action instead of the alternative.

Comparing alternatives: While not required for an EA, a comparison of alternatives could be advantageous as it can help foster more informed decision making. The results of such an investigation can provide the decision maker with factors for better decision making. Table 7.12 provides a hypothetical comparison of three potential courses of action for obtaining a source of steam and electric power (e.g., cogeneration):

1. Construct a small cogeneration plant.
2. Renovate an existing plant.
3. Take no action with respect to replacing an existing cogeneration plant.

Instead of using simple responses such as "Yes" or "No," many practitioners find it more informative to use quantitative or short descriptive phrases. For example, instead of indicating "Yes" to the question of land disturbance, a response might state, "Grading of 0.75 acres." Instead of indicating "Yes" to the question of demolition, a response might state "Limited to three office buildings constructed prior to 1950."

7.5.4 Environmental impact (environmental consequences) section

It is recommended that the environmental impacts section, often termed the environmental consequences section, of an EA begin with an evaluation of the potentially most important issues or impacts, ending with those of least importance. The order used for describing the impacts to environmental resources should ideally follow the same sequence as that used in describing the resources in the affected environment section.

Considerable misunderstanding exists regarding the types of environmental impacts that are possible and how each type of impact should be addressed. The CEQ in the regulations defines three broad categories of environmental impacts that must be considered under NEPA:

- Direct: caused by the action and occur at the same time and place (40 CFR § 1508.8).
- Indirect: caused by the action but later in time or farther removed in distance, but are still reasonably foreseeable. Indirect effects may include growth inducing effects and other effects related to induced changes in the pattern of land use, population density or growth rate, and related effects on air and water and other natural systems, including ecosystems (40 CFR § 1508.8).
- Cumulative: impacts resulting from the incremental impact of the action when added to other past, present, and reasonably foreseeable future actions regardless of what agency (federal or nonfederal) or person undertakes such other actions. Cumulative impacts can result from individually minor but collectively significant actions taking place over a period of time (40 CFR § 1508.7).

It is not always easy to attribute every possible impact to one of the three categories. The line between direct and indirect (sometimes termed "secondary") impacts can be especially blurry. Forest clearing within a right-of-way would clearly be a direct impact in an EA (or EIS) for a road construction project, and forest clearing as a result of induced commercial development around a new interchange would clearly be an indirect impact. But would forest clearing on rented private land off of the right-of-way to accommodate construction trailers be direct or indirect? However, precisely distinguishing between the categories of impacts is less important than ensuring that no impacts falling into any one or more of the categories are overlooked. Fortunately, the regulations do not require preparers to NEPA documents to classify each impact—they instead focus on ensuring that all relevant impacts are considered, irrespective of classification.

Most EAs (and EISs) present a combined discussion of direct and indirect impacts for each resource topic. That approach meets the objectives of NEPA while obviating the need to decide which impacts are direct and which are indirect. Many EAs (and most EISs) discuss cumulative impacts separately, however. This is at least partially due to the fact that cumulative impacts are considerably more difficult to intuitively understand and analyze and that the first meaningful CEQ guidance on cumulative impact assessment was only issued in 1997 (more than 25 years after implementation of NEPA). Especially in shorter EAs, an integrated discussion of direct, indirect, and cumulative may be more effective (and easier to read) than separate discussions. The integrated discussion should however clearly indicate that direct, indirect, and cumulative impacts were considered.

The discussion of direct and indirect impacts is not usually segregated into separate sections. But because of impact complexity and the special analytical methodologies that are sometimes necessary, the cumulative impact analysis is sometimes delegated to a separate section. However, there is no requirement to analyze cumulative impacts in a separate section. In fact, for most simple EAs, there are advantages to integrating the discussion of direct, indirect, and cumulative impacts. These advantages include increased document simplicity and ensuring a holistic discussion of all categories of impacts. Integrated discussions should always, however, specifically address cumulative impacts by name.

The impacts, or effects, analyzed in EAs must include a discussion of impacts that are expected to result from:

1. The conduct of the proposed action itself or any of the alternatives (direct impacts).
2. Activities that are not a part of the proposed action or any of the alternatives but are reasonably foreseeable consequences of an agency conducting the proposed action or alternatives (indirect impacts).

This chapter must discuss these impacts in each of the following contexts:

1. Viewing the direct and indirect impacts of the proposed action and alternatives as if it were the only activity being conducted (individual impacts).
2. Viewing the direct and indirect impacts in the context of all other activities (human and natural) that are occurring in the affected environment and impacting the resources being affected by the proposed action and alternatives (cumulative impacts).

Analysis of cumulative impacts—The CEQ directs in the regulations that cumulative impacts be considered in addition to direct and indirect effects in reaching a determination regarding significance (§1508.27[b][7]). As cumulative impact assessments have increasingly become the focus of litigation in recent years, the importance of performing an adequate analysis has correspondingly risen. Notwithstanding, the analysis of cumulative impacts is often either neglected or is not afforded adequate attention, especially in EAs.

For example, one study reviewed cumulative impact considerations in 30 different EAs prepared for a variety of different projects throughout the United States. Out of the 30 EAs, the term "cumulative impact" was mentioned in only 14 (47%) of the assessments. Moreover, only 7 (23%) of the EAs discussed cumulative impact in the environmental consequences section of the EA. When cumulative impacts were mentioned, they were typically addressed in a qualitative manner and lacked a clear delineation of the spatial and temporal study boundaries. The study's authors concluded that neither the discussion of cumulative impact considerations nor the thoroughness of the analysis was generally sufficient to support informed decision making.[15]

The findings of this study are consistent with the conclusions reached in a similar study of 89 EAs. Out of the 89 EAs, only 35 (39%) mentioned the term "cumulative impact," with eight of these assessments stating that there were no cumulative impacts but providing no evidence to support such a conclusion. Five EAs identified potential cumulative impacts but concluded they were nonsignificant without presenting an analysis to support such a conclusion.[16]

In some EAs, the issue of cumulative impacts may be the single most important factor in determining whether a FONSI can be supported. If EAs are to be credibly used in reaching significance determinations, the issue of cumulative impacts must be addressed in a more rigorous and systematic manner.

Describing impacts—Agencies must make a concerted effort to "... disclose and discuss ... all major points of view" (§1502.9[a]). Graphics can be particularly useful in conveying pertinent information to the decision maker and public.

Instead of simply stating that a resource "would be impacted," the analysis must indicate how and to what extent environmental resources would be affected. For instance, it is much more informative to describe how much and to what degree a critical habitat would be perturbed than to simply state (as is sometimes the case) that the "... habitat would be impacted." Where both possible and central to the issues addressed in an EA, impacts should be quantified using an appropriate unit of measurement. Potentially affected resources or populations should be clearly described. The time and period over which the impact would occur should be indicated, as should its likelihood or probability. Although probability is ideally expressed as a number between 0 (no likelihood) and 1 (certainty), neat mathematical expression of probability is rarely possible in environmental impact assessment, thereby relegating assessors to verbal descriptions of likelihood.

It is usually not necessary to categorize each impact as "short-term" or "long-term," or "direct" or "indirect." Selected topics that necessitate special elaboration are described in the following sections.

EAs, especially brief EAs for simple proposals, do not generally have to contain as much quantitative detail as do EISs. However, for an EA to avoid the need for an EIS, it must contain adequate detail to support a FONSI. An EA, regardless of length, must contain the technical detail and quantitative precision needed to defend the FONSI. Brevity and simplicity are desirable traits for an EA only if the EA provides the technical basis for a well-grounded FONSI.

Site-specific analyses: Case law indicates that the impact analysis must be site-specific. Lack of data in databases maintained by federal or state agencies on sensitive or endangered species at a site does not imply that the site has been studied and found to lack such species. Indeed, most state databases on locations of threatened and endangered species only include observations from scientific surveys conducted on scattered parcels of land, usually public lands that have been the subject of recent NEPA and federal and state environmental permitting efforts. Almost any EA involving some type of construction that could affect naturally vegetated areas should include an evaluation, usually including a site visit, by a biologist. Similar investigations might be required where the proposal could affect other environmental resources, e.g., cultural resources, as well.

Conformance with regulatory standards: Conformance with regulatory standards does not necessarily ensure that an action would not significantly affect an environmental resource. A statement such as, "The activity would be conducted in accordance with all applicable regulatory requirements ..." should not be used as the sole or even primary source of evidence of nonsignificance. An action could comply with all applicable laws and regulations and still result in a significant environmental impact.

For example, an impact to a wetland area that qualifies for an exemption or authorization under a general permit to Section 404 of the Clean Water Act is not necessarily nonsignificant in the context of NEPA. Certain impacts to wetlands, such as atmospheric emissions, noise, and even some types of draining, might not even be regulated under Clean Water Permitting Programs. Some wetlands lacking surface connections to stream systems have been designated as "isolated" wetlands by the U.S. Army Corps of Engineers and have been determined to be exempt from Clean Water Act regulation. However, impacts to some isolated wetlands can have substantial ecological or hydrological consequences.

As another example, shipment of radioactive waste might comply with all applicable laws and regulations. Yet, the fact that such a shipment would comply with all regulatory standards does not necessarily guarantee that a significant impact could not occur (e.g., the transportation truck has an accident, or the cargo ship strikes another ship and sinks). A good impact analysis would consider the likelihood and possible consequences of such accidents.

Conformity determinations with air quality standards: Under the NEPA regulations, a determination of significance requires considering "whether the action threatens a violation of federal, state, or local law, or requirements imposed for the protection of the environment" (§1508.27[b][10]). A directive by the EPA establishes specific direction for reviewing potential violations of existing State Implementation Plan (SIP) conformity criteria under the National Ambient Air Quality Standards (NAAQS).[17] Specifically, this directive provides a mechanism for reviewing and enforcing criteria pollutant standards (i.e., carbon monoxide, lead, nitrogen dioxide, ozone, particulate matter, and sulfur dioxide).

Pursuant to EPA's direction, conformity determinations for proposed actions are to be integrated with the NEPA process. EPA's directive defines exemptions and the specific circumstances under which conformity must be addressed in an NEPA document. Where applicable, an EA might need to consider direct and indirect emissions in evaluating conformity with SIP criteria pollutant standards.

Mitigation measures—The CEQ defines mitigation in the regulations as:

- Avoiding the impact altogether by not taking a certain action or parts of an action.
- Minimizing impacts by limiting the degree or magnitude of the action and its implementation.
- Rectifying the impact by repairing, rehabilitating, or restoring the affected environment.
- Reducing or eliminating the impact over time by preservation and maintenance operations during the life of the action.
- Compensating for the impact by replacing or providing substitute resources or environments.

Identification of mitigation measures has been a key part of the EIS process since the initial implementation of NEPA. Mitigation has until recently played a smaller role in the EA and FONSI process. In its answer to Question 39 of the *40 most asked NEPA questions*, the CEQ indicates that an EA might still identify suitable mitigation for a proposal even though the action has been found to have no significant impact. However, in its answer to Question 40, CEQ places limitations on the ability of agencies to issue a FONSI, and hence avoid an EIS, by implementing mitigation as a means to reduce impacts to nonsignificance.

The CEQ in the regulations originally contemplated that agencies would use a broad approach in defining significance and would not rely on the possibility of mitigation as an excuse to avoid the EIS requirement." A number of subsequent court cases have ruled, however, that mitigation may in fact be used to reduce significant impacts to the point that an EIS is no longer needed. FONSIs issued on the basis of impact reduction through mitigation are termed "mitigated FONSIs." The possibility of using mitigation to avoid the time and expense of the EIS process serves to motivate agencies to implement reasonable mitigation.

In early 2010, the CEQ issued a draft memorandum that further cements mitigation as a way to qualify for a FONSI. Although the memorandum addresses mitigation in the context of both EISs and EAs, it includes substantial language on the use of mitigation in EAs and FONSIs. For EAs, it emphasizes that mitigation measures must be clearly explained in the EA and expressed as a commitment in the FONSI. It directs agencies proposing mitigation in an EIS or EA to establish a program of implementation and monitoring. The memorandum states that agencies should take action to address mitigation failures. It states that if an agency proposing an action covered by a mitigation FONSI finds that it cannot effectively prevent mitigation failures, it should prepare an EIS. Of particular note, the memorandum emphasizes the use of public involvement throughout the processes of planning and implementing mitigation. It does not imply that the standard for public involvement should be lower for an EA than for an EIS.

If an agency plans to issue a mitigated FONSI, it must include a description of proposed mitigation measures in its EA. The proposed mitigation is frequently included in the environmental impact section of an EA. Sometimes, mitigation measures are described and incorporated as an integral element of the proposed action. If so, analysis of their effectiveness in mitigating potential impacts should still be discussed in the environmental impact section of the EA. It may be possible to cite one or more stand-alone mitigation plans prepared for specific resources, e.g., a wetland mitigation plan or a noise abatement plan. Such plans are commonly attached as appendices. If so, the EA can merely summarize the mitigation measures and evaluate their effectiveness and likelihood of success, and refer the reader to the referenced plan document for design details.

Mitigation measures should be described as specifically as possible. Equally important, the practicality of implementation and likelihood of success should be evaluated. Mitigation measures that cannot be practicably implemented will most likely never be implemented and hence not be effective. Decision makers need at least some understanding of the

Table 7.13 Five elements essential in mitigation measures

1. Specific and detailed measures, including an explanation of how those measures would reduce impacts
2. Specific schedule
3. Appropriate funding
4. Likelihood of success and measurable performance criteria
5. Assignment of responsibility for implementation

risk of failure of mitigation measures before issuing a FONSI that relies on those measures. Statements such as, "A mitigation plan will be developed later in cooperation with the EPA," do not provide sufficient evidence for a decision maker to determine whether significant impacts can be mitigated to the point of nonsignificance. A well-planned and defensible mitigation plan should contain five essential ingredients depicted in Table 7.13.

Proposed measures for monitoring and documenting the success of mitigation measures should also be described. The U.S. Army Corps of Engineers and EPA have become increasingly concerned over the lack of maintenance and monitoring of wetland mitigation projects and recently issued a rule requiring stricter assurances of mitigation project success. Mitigation must be a plan for action, not merely a concept outlined on paper.

Agencies sometimes debate whether actions that avoid or reduce environmental impact should be counted as "mitigation" if required by permits or statutes outside of NEPA. For example, most municipalities require that construction activities follow "best management practices" (BMPs) to reduce or prevent soil erosion and sedimentation of streams and waterways. Examples of common BMPs include seeding exposed soils and constructing basins to trap and detain runoff that would otherwise have entered streams. These practices certainly meet the CEQ definition of mitigation, despite their routine application outside of the context of NEPA. Any argument as to whether the practices are "official" mitigation measures is largely academic; what is important is that the agency actually implements them. Whether inherent in the impact analysis or part of proposed mitigation, the EA should evaluate how such BMPs might reduce impacts. In either case, the EA should describe each BMP and its effectiveness and likelihood of success in detail and not rely on vague generalities such as "best management practices would be implemented as required by statutes and permit conditions."

Comparing impacts among alternatives—While not required for an EA, a comparison of environmental impacts that would result from pursuing reasonable alternatives can be advantageous as it can assist the agency in determining an optimum course of action. The results can provide the decision maker with factors that foster more informed decision making.

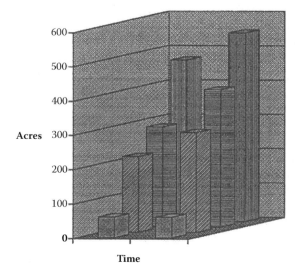

■ No Action	60	60
▨ Proposed Action	220	290
■ Alternative 1	290	400
■ Alternative 2	470	550

Figure 7.2 Acres of disturbed habitat.

Agencies must make a concerted effort to "... disclose and discuss .. .all major points of view" (§1502.9[a]). Graphics can be particularly useful in conveying pertinent information to the decision maker and public. For example, the bar chart shown in Figure 7.2 shows the amount of habitat that would be disturbed with respect to four different courses of action. The first row of bars shows the habitat disturbance that would result during the construction phase of a proposed military training facility. The second row of bars depicts the disturbance that would result during the operational phase.

7.5.5 *Listing permits, approvals, and regulatory requirements*

Listing applicable permits and regulatory requirements helps agencies in identifying and planning for future actions, as well as in assessing potentially significant impacts (i.e., §1508.27[b[10]). A section briefly (see Section 8.0 in Table 7.4) listing applicable federal, state, local, and tribal regulatory approvals and requirements can assist the decision maker in assessing the potential for significant impacts. The list can also facilitate agency planning by identifying future requirements during the early planning stage.

Even though NEPA is a federal statute, the list usually includes state and local environmental regulations. Many states and some localities have environmental regulations that closely parallel federal environmental regulations. Common examples include state regulations addressing clean air and water, endangered species, and historical preservation. Some state and local regulations may have more restrictive permitting thresholds, e.g., maximum pollutant concentrations in water, than corresponding federal standards. The interpretation of what constitutes an "environmental" regulation is usually interpreted broadly; for example, local zoning approvals and building permits are generally relevant to an environmental analysis.

As stated in the previous section, a common misperception is that compliance with permits and other regulatory requirements implies insignificance of impacts. This is emphatically not true! Not every type of environmental impact is the subject of a specific statute. For example, many states do not regulate impacts to species identified by that state as rare, threatened, or endangered in the state. Few states regulate forest clearing, and many states do not regulate wetland impacts (leaving as unregulated any wetlands not federally regulated under the Clean Water Act). NEPA is a comprehensive environmental planning statute that calls for "integrated use of the natural and social sciences and the environmental design arts." Furthermore, keeping the magnitude of impacts below compliance thresholds established by statutes does not necessary imply that the impacts are not significant. Thresholds established by statute are often the result of political as well as scientific considerations; they are often liberalized as a compromise to attain the votes needed for passage.

7.5.6 Listing of agencies and people consulted

The CEQ guidelines specifically noted in 40 CFR 1508.9 that EAs should include a listing of agencies and people consulted. One of the purposes of this requirement is to publicly indicate the degree to which agencies have conducted a thorough and interdisciplinary process. Consultation is also an important regulatory mandate that facilitates a more thorough analysis and is useful in identifying related permitting requirements.

Considering the CEQ's lack of specificity on most other elements of EA preparation, it is clear that the CEQ considers this list to be an important component of any EA. Citing agencies and people consulted is helpful in demonstrating that appropriate experts were consulted and that an interdisciplinary analysis was prepared. It can also demonstrate that the assessment was coordinated with other agencies having jurisdiction or special expertise with respect to the environmental issues. While it is almost impossible to write an EIS or long EA without the input of an interdisciplinary team of experts, it might be possible for a single individual to write a short EA without input or assistance from appropriate experts. While it is generally acceptable for a single individual to be the sole author of an EA, such a single-author EA should clearly demonstrate that that author solicited input from appropriate regulatory and technical experts.

The list should include names of consulted individuals as well as agency or affiliation names. The individual's title and any qualifications demonstrating the appropriateness of the individual should be stated. Merely stating that an agency was consulted does not demonstrate that an appropriate expert was consulted. In a sense, the individuals consulted function as coauthors of the EA, in that they are sources of data or analyses contributing to the utility of the document.

The list should not be limited just to people consulted as part of formal consultations performed in compliance with regulatory requirements under statutes such as the Endangered Species Act, Fish and Wildlife Coordination Act, or National Historic Preservation Act. While those consultations are important and should be listed, the list should include less formal communications with various agencies, commissions, boards, and experts. Some examples of possibly useful individuals for consultation in an EA include:

- Federal agencies
- State agencies
- Local agencies and offices
- Regional technical experts, often affiliated with universities
- Regional commissions, such as those established for certain watersheds
- Regional nonprofit organizations.

Beyond consultations performed for regulatory compliance, the choice of consultation subjects is at the discretion of the author(s) of the EA. The purpose of consultations is to gather information and informed opinions; the need for consulting a given subject is driven by the relevance of the information that subject might provide. Communications may be in person, by telephone, by mail, or even by e-mail. Ideally, EA preparers should prepare written summaries of each communication to include in the administrative record. Summaries containing key information may be appendicized to the EA. As a matter of courtesy, EA preparers should inform consulted individuals that their names will be included in a published document.

7.5.7 List of preparers

The CEQ guidance does not prescribe including a list of preparers for EAs as it does for EISs. Listing preparers can, however, help the agency in demonstrating that it performed an interdisciplinary analysis (§1501.2[a]) and can help in substantiating the analysis and documenting its administrative record. As for an EIS, information on authorship and qualifications adds credibility to the analyses. In a sense, the authors of the EA should be counted among the "people consulted" contributing their expertise to the analysis, even if they are employees or contractors of the agency. Additionally, identifying authorship promotes quality by instilling pride in the analysis and helps professionals build an easily verified record of experience. An analogy is the credit list included at the end of movies and television shows.

For EISs, the CEQ guidance calls for listing the names and qualifications of each author and preparer of significant background papers. Many EISs and longer EAs are authored primarily by contractors with oversight and review by agency personnel. Preparer lists for such documents should list each contractor staff member contributing to the document, not merely identify the contractor name or contractor project manager. Agency subject matter experts actively involved in the analysis and writing process should also be included. Agency experts who lend their expertise in the form of a technical review should be included. Personnel who serve only in a supervisory or managerial role are best included only if they served a clearly defined role in the subject project. Often overlooked are graphics designers and technical editors; their expertise is critical to the success of the project and should be formally acknowledged.

Also commonly overlooked are preparers of key background analyses and papers. Preparers of published sources referenced in the text need not be listed; the reference list provides adequate acknowledgement. But preparers of unpublished papers key to the analysis in the EA but not publicly available and citable need to be acknowledged. Examples might include unpublished biota surveys, wetland delineations, and historic and archaeological surveys. Professionals who perform modeling, simulations, or other technical analyses that are relevant should be included even if the results were never presented in a report format. It does not matter whether every "preparer" in the list was directly involved in writing the EA text.

Many shorter EAs are prepared by only one or two staff personnel. A full blown "list" may not be necessary, but the responsible personnel should still be identified. Names can be noted on the cover, inside cover page, a cover letter, or by footnotes or endnotes instead of an actual list. Even EAs written in their entirety by a single author might rely on the analytical contributions of several other individuals.

References

1. Doub, J. Peyton and C.H. Eccleston, A Systematic Tool for Determining the Need for Specialized Expertise in NEPA, *Environmental Practice*, 5(4): 288–9.
2. Council on Environmental Quality, *Considering Cumulative Effects Under the National Environmental Policy Act*, January 1997.
3. NEPA Forty Most Asked Questions.
4. Council on Environmental Quality. Memorandum for heads of federal departments and agencies dated February 18, 2010, from Nancy H. Sutley, chair. Subject: *Draft Guidance for NEPA Mitigation and Monitoring*.
5. U.S. Army Corps of Engineers and U.S. Environmental Protection Agency, April 10, 2008. *Compensatory Mitigation for Losses of Aquatic Resources; Final Rule*, 73 FR 19594-19705.

chapter eight

Assessing significance

> A man who picks up a cat by the tail learns something he can learn in no other way.
>
> —**Mark Twain**

8.1 Introduction

Assessing the significance of potential environmental impacts is both a complex problem and central to the National Environmental Policy Act (NEPA) goal of protecting environmental quality. It is especially central to the EA (environmental assessment), the subject of this book. As indicated in previous chapters, a defensible and systematic procedure for determining the threshold of significance is also crucial as it dictates when an EA can suffice as adequate environmental impact analysis or when an environmental impact statement (EIS) is required. It also drives the depth of study needed for each issue, whether within an EA or EIS. Potential significance also plays a role, although not an exclusive role, in determining the likelihood of an impact being raised during scoping and the number and complexity of substantive comments raised following publication of an EA or draft EIS. This chapter examines the concept of significance and provides the reader with a systematic framework for making such determinations. It also describes a systematic procedure developed by Frederic March for assessing significance.[1] A more detailed review of significance and its determination can be found in Eccleston's companion text, *NEPA and Environmental Planning*.

8.2 Definitions and use of the term "significance"

The concept of significance determines whether:

- An action is considered "major."
- An action can be categorically excluded.
- An EA should be prepared for a proposal.
- A finding of no significant impact (FONSI) can be issued.
- An EIS needs to be prepared.

- The degree of treatment (range and depth) that should be devoted to a particular significant issue or impact addressed in an EIS (i.e., sliding-scale approach).

The concept of significance in the context of NEPA is not identical to the concept of significance in statistics. Statistical significance is a mathematical concept using calculations to compare quantitative differences between numerical data sets reflecting multiple conditions or experimental treatments. Scientists rely on statistical calculation procedures to test hypotheses regarding the presence of meaningful differences between measured outcomes of experiments or suspected trends in quantitative observations. Statistics finds applications in most natural and social sciences. With NEPA the term "significance" is determined more judgmentally, requiring the careful weighing of both quantitative and qualitative evidence by multidisciplinary teams of experts. Statistical significance is expressed using numerical values whereas NEPA significance is usually expressed as either "significant" or "not significant" (or "insignificant"), although some NEPA documents do acknowledge broad classes of significance. Statistical significance can be used in any context where there are differing sets of quantitative data, whereas NEPA significance is limited to assessing impacts of actions on baseline conditions.

8.2.1 Potential significance should determine the depth of analysis

Chapter 6 and Chapter 7 of this text emphasize that the depth of analysis for an environmental issue in an EA, and consequently the length of discussion in an EA, should use a "sliding-scale" approach focusing on the potential for significant environmental impacts. The following selected citations explicitly indicate the pivotal importance and central role that significance plays:

- "All agencies of the federal government shall ... include in every recommendation or report on proposals for legislation and other major federal actions *significantly* affecting the quality of the human environment a detailed statement" (NEPA, 102[2][A], emphasis added).
- "Most important, NEPA documents must concentrate on the issues that are truly *significant* to the action in question rather than amassing needless detail" (§1500.1[b], emphasis added).
- "Using a *finding of no significant impact* when an action not otherwise excluded will not have a *significant effect* on the human environment and is therefore exempt from requirements to prepare an environmental impact statement" (§1500.4[q] & §1500.5[l], emphasis added).

The terms "significant effect," "significant issue," and "significant impact" are used in reinforcing the intent of the above citations (§1501.1(d), §1501.7, §1501.7(a)(2), §1501.7(a)(3), §1502.1, §1502.2(b), §1505.1(b), §1508.4).

8.2.2 Need for a systematic approach

With respect to the problem of determining significance, NEPA practice can fall short in two ways. Specifically, the practitioner may fail to:

1. Provide adequate evidence that impacts are not significant, resulting in challenges to FONSIs or to the adequacy analytical rigor in EISs.
2. Identify what is truly significant and what constitutes needless detail resulting in NEPA documents (whether EAs or EISs) of excessive cost and delay in implementation of federal programs.

The following section advances a systematic approach, referred to as the significance assessment tests, for determining significance, based on strict adherence to the following factors:

1. "Significantly" is understood in the strict sense defined by §1508.27.
2. Application of rigorous and systematic tests for assessing significance, based on both letter of the law and spirit of the NEPA act.

The following discussion focuses primarily on EA preparation, but much of it is applicable to EIS preparation and other elements of NEPA compliance.

8.3 General procedure for determining significance

The potential significance of an action is defined in the NEPA regulations (regulations) as having two separate but closely related aspects: context and intensity.

As outlined by March, a close examination of the intensity factors provided in 40 CFR 1508.27 of the Council of Environmental Quality's NEPA regulation, yields seven specific tests, depicted in Table 8.1, that can be systematically applied in determining the significance of any given action.

Thus the significance assessment test or SAP actually is a set of seven interrelated tests. It is recommended that the decision maker use this procedure in reaching each and every FONSI determination. It is furthermore recommended that the decision maker document the consideration given to each of the seven tests in the agency's administrative record, specific

Table 8.1 Seven tests for assessing significance

1. **Receptor Test:** Identifying the presence or absence of certain environmental receptors, and then considering the impacts only on those receptors that are potentially affected (§1508.27[b][2,3,8,9]).
2. **Activity Test:** Identifying the presence or absence of certain activities and then considering the impacts only of those activities that are likely to affect the above receptors.
3. **Regulatory Compliance Test:** Determining whether the impacts include a threatened violation of any or local law, regulation, or requirement respecting the environment. In most cases, these laws, regulations, and requirements provide threshold tests for compliance. This test covers a wide range of effects including health and safety as well as other regulations protecting air, water, land, and ecological resources (§1508.27[b][10]).
4. **Risk/Uncertainty Test:** Determining the degree to which the possible effects on the environment are highly uncertain or involve unique or unknown risks. This requirement includes risks resulting from natural hazards and from accidents (§1508.27[b][5]).
5. **Cumulative Test:** Determining whether the action is related to other actions with individually nonsignificant but cumulatively significant impacts (§1508.27[b][7]).
6. **Precedence Test:** Determining the degree to which the action might establish a precedent for future actions with significant effects or represents a decision in principle about a future consideration (§1508.27[b][6]).
7. **Controversy Test:** Determining the degree to which the effects on the quality of the human environment are likely to be highly controversial (§1508.27[b][4]).

metrics (criteria and threshold values) should be established when possible and practicable for assessing these intensity factors to determine whether a significant impact would be incurred. Of course, not all types of potential impacts (for example, aesthetic impacts to visual resources) lend themselves to precise or even rough quantification. But NEPA practitioners should be able to apply qualitative considerations to these categories of impacts following the same general thought process applied to quantifiable impacts (for example, impacts related to water withdrawals or discharges). Remember that significance determination in NEPA does not always depend on the mathematical rigor used in assessing statistical significance.

8.3.1 *Description and basis of the tests*

To qualify for a FONSI, a proposed federal action must pass all of the tests exhibited in Table 8.1 (i.e., passing each test means qualifying as nonsignificant in relation to that test.).

First two tests—The first two tests shown in Table 8.1 merely establish certain receptors and activities that are considered in the subsequent tests. While the CEQ regulations do not identify activities *per se*, they instruct agencies to develop procedures, including lists of activities that can be categorically excluded, or that would normally require preparation of an EA or an EIS.

Regulatory Compliance Test—After identifying potential environmental receptors and activities, an examination of compliance with applicable federal, state, and local codes and regulations (see Table 8.1, Test No. 3) is an excellent initial screen. As a practical matter, many small actions that pass this test will often pass the subsequent tests as well. However, a common error has been to assume that meeting this factor alone is sufficient to establish nonsignificance. Not all types of potentially significant environmental impacts are regulated, and for those that are, the regulatory compliance standards may have been limited by political concerns unrelated to true thresholds of environmental significance.

Risk/Uncertainty Test—This test (see Table 8.1, Test No. 4) is applied in cases where the action could create unavoidable risks. Such risks might involve those associated with industrial operations, a treatment plant for hazardous or radioactive wastes, or projects in areas subject to natural hazards such as earthquakes, tornadoes, and hurricanes. As a practical matter, socially acceptable risks are often specified in regulations, design codes, and standards of practice, so if the action passes the Regulatory Compliance Test, it will frequently pass the risk test. The risk/uncertainty issue can also be triggered by considerations related to the Cumulative Test, the Precedence Test, and the Controversy Test.

Cumulative and Precedence Tests—The Cumulative and Precedence Tests (see Table 8.1, Tests No. 5 and No. 6) least lend themselves to objective criteria as the methodological basis for predicting future effects in a complex setting is often subjective. As a result, when these criteria are at issue, they will likely be associated with the risk/uncertainty and controversy factors.

Controversy Test—The Controversy Test (see Table 8.1, Test No. 7) might reopen the Risk/Uncertainty Test, as many controversies center around the risk issue. Controversy is less likely to be associated with regulatory compliance (although it is possible) and more likely to relate to the action's ability to pass the Cumulative and Precedence Tests described above. Scientific controversy levels for many issues (for example, climate change) are as much a factor of public perception as they are of demonstrable environmental significance.

8.3.2 Assessing the seven tests of significance

The balance of this chapter provides a specific methodology for assessing each of the seven tests described in Table 8.1.

Receptor Test—The Receptor Test (see Table 8.1) is based on the fact that certain environmental receptors (shown in bold, Table 8.2) that can be used in testing a given action for significance are specifically prescribed by §1508.27(b)(2), (3), (8), and (9).

Definition of environmental receptor: The factors cited in Table 8.2 form the core of a checklist together with a number of other receptors subject to potentially significant effects. These include receptors cited elsewhere in CEQ's regulation, receptors associated with executive orders (EOs), and receptors associated with the specific context in which an action is proposed. The term "environmental resource" is commonly used as a synonym for the term "environmental receptors." As used herein, the term environmental receptor means:

1. Any feature of the environment, natural or manmade.
2. Any renewable or depletable resource.
3. Any measure or statistic describing a feature of the environment, such as population density or monitored air quality, that might be impacted by a proposed action.

Checklist of receptors: Developing and applying a checklist of receptors is a preliminary step that enables the practitioner to document the comprehensive consideration given to the assessment of potential effects. For example, the U.S. Air Force has used a checklist known as Form 813—Preliminary Environmental Impact Analysis. Such checklists can be helpful in determining whether the action qualifies for a categorical exclusion, is likely to qualify for a FONSI if an EA is written, or is likely to require an EIS.

Table 8.2 The receptor test can be assessed in terms of certain significance factors prescribed in §1508.27(b)(2),(3),(8), and (9)

---
1. The degree to which the proposed action affects **public health** or **safety**.
2. Unique characteristics of the geographic area such as proximity to **historic** or **cultural resources, park lands, prime farmlands, wetlands, wild** and **scenic rivers,** or **ecologically critical areas.**
3. The degree to which the action may adversely affect **districts, sites, highways, structures,** or objects listed in the **National Register of Historic Places** or may cause loss or destruction of significant **scientific, cultural,** or **historical resources.**
4. The degree to which the action may adversely affect an **endangered species** or its **habitat** that has been determined to be critical under the Endangered Species Act of 1973.

Chapter eight: Assessing significance

Checklists should contain at least all of the items in the regulations, EOs, and other directives requiring that the receptor somehow be considered. A good classification scheme is logical and hierarchical so that, in principle, it can be expanded to whatever level of detail is needed.

Table 8.3 provides a receptor checklist that systematically encompasses every receptor mentioned in the regulations and certain other environmental laws, regulations, or requirements. While only some of these receptors are explicitly stated in the factors for significance (§1508.27), all of the items listed in Table 8.3 should be considered because a significant impact on any one receptor can render the entire action significant.

Procedure for applying receptor test: Recommended practice applying the Receptor Test is indicated in Table 8.4.

Activity Test—Table 8.5 lists a number of activities suitable for inclusion on a checklist. The table indicates the federal law, regulation, or requirement that corresponds to each activity. A thorough checklist exercise screens receptors and activities together and takes note of appropriate cause and effects relationships.

The recommended practice for applying the Activity Test is indicated in Table 8.6.

Regulatory Compliance Test—The following significance factor lends itself to establishing a number of "threshold" tests of whether an impact is significant (§1508.27[b][10]):

> Whether the action threatens a violation of federal, state, or local law or requirement imposed for the protection of the environment.

Tables 8.3 and 8.5 provide a partial list of federal laws, regulations, and requirements associated with respective receptors and activities. A comprehensive guide would cite all applicable regulations and highlight those sections that prescribe standards or other tests of compliance. In addition to the federal list, the practitioner should be aware of applicable state and local laws, regulations, and ordinances.

Table 8.7 shows the recommended practice for applying the Regulatory Compliance Test.

As indicated earlier, a common error in NEPA practice is to assume that an action simply meeting the Regulatory Compliance Test is not significant. Therefore, it must be stressed that regulatory compliance is only one of seven distinct tests that must be applied in determining significance. For example, the following circumstances could result in significant impacts in spite of passing the regulatory compliance:

> The applicable federal, state and local laws, regulations, and requirements are arguably deficient and do not prevent significant impacts per the other tests.

Table 8.3 Receptor-related requirements

Receptor	CEQ citation	Other federal laws, regs., or reqs.
Air		
Air Resources	§1508.8(b)	CAA, Pollution Prevention Act
Nonattainment Area		CAA Conformity (40 CFR 6, 51 & 93)
Class I Control Region		CAA Conformity (40 CFR 6, 51 & 93)
Radioactive Exposures		CAA-NESHAPS (40 CFR 61)
Hazardous Exposures		CAA-ESHAPS (40 CFR 61)
Acid Rain		CAA
Ozone Layer		CAA
Odor		Nuisance Case Law
Visibility		CAA
Navigable Air Space		14 CFR 77.13a
Water		
Water Resources	§1505.8(b)	CWA, SDWA, Pollution Prevention Act
Wild and Scenic Rivers	1508.27(b)(3), 1506.8	CEQ Memoranda, FR Vol. 45, No. 175, 9/8/80
Special Sources (Ground)		Safe Drinking Water Act, 40 CFR 149
Navigable Waters		Clean Water Act, Oil Pollution Act
Land		
Land Use	§1502.16(c), §1508.8(b)	
Park Lands	§1508.27(b)(3)	
Prime Farmlands	§1508.27(b)(3)	CEQ Memoranda, FR Vol. 45, No. 175, 9/8/80
Wetlands	§1508.27(b)(3)	Executive Order 11990
Floodplains		Executive Order 11988
Coastal Zone		Coastal Zone Management Act, 15 CFR 930
Soils/Geologic		RCRA/CERCLA
National Priority Sites		CERCLA

(continued)

Table 8.3 Receptor-related requirements (continued)

Receptor	CEQ citation	Other federal laws, regs., or reqs.
Human health/safety		
Public Health & Safety	§1508.27(b)(2)	NEPA—Sec. 2, 101(b)(2,3), 101(c)
Radioactive Exposures		CAA-NESHAPS (40 CFR 61), Atomic Energy Act, Nuclear Reg. Comm., 10 CFR 20—Rad. Stds.
Hazardous Exposures		CAA-NESHAPS (40 CFR 61) OSHA, CWA, SDWA, TSCA, RCRA, CERCLA, EPCRA, Pollution Prevention Act, Hazardous Materials Transportation Act
Species/habitat		
Endangered Spec./Hab.	1508.27(b)(9)	NEPA—Sec. 2, 102(H)
Ecologically Critical Areas	1508.27(b)(3),1508.25	Endangered Species Act
Ecosystems	1508.8(b),1508.25	Fish and Wildlife Coordination Act
Ecological Resources	1508.8(b)), 1507.2(e)	Marine Mammal Protection Act
		Marine Protection, Research & Sanctuaries Act,
Biosphere		NEPA—Sec. 2
Wilderness	1506.8(b)(2)(ii)	Wilderness Act
		Also see: Environmental Protection Agency: Habitat Evaluation, Guidance for the Review of Environmental Impact Assessment Documents, January 1993.
Social/economic		
Social	1508.8(b)	NEPA 101(a), 101(b)(4)
Economic/Standard of Living	1508.8(b)	NEPA 101(a), 101(b)(5)

(continued)

Table 8.3 Receptor-related requirements (continued)

Receptor	CEQ citation	Other federal laws, regs., or reqs.
Aesthetic	1508.8(b)	NEPA 101(b)(2)
Growth	1508.8(b)	NEPA 101(a)
Urban Quality	1502.16(g)	NEPA 101(a)
Built Environment	1502.16(g)	
Population Density	1508.8(b)	NEPA 101(a)
American Indians	1502.16(c)	American Indian Religious Freedom Act
Minority/Low-Income		Executive Order 12898—Environmental Justice
Life's Amenities		NEPA 101(b)(5)
Historical/cultural/scientific/national heritage		
Historical/Cultural	1508.27(b)(3,8), 1502.25, 1508.8(b), 1502.16(g)	NEPA 101(a), 101(b)(2,4), Nat. Hist. Preservation Act of 1966, Historic Sites Act of 1935, Antiquities Act of 1906, Natl. Regist. of Hist. Places
Archaeological		Archaeological Recovery Act of 1960, Archaeol. Resources Protection Act, Archaeol. and Historic Preservation Act of 1974
Scientific	1508.27(b)(8)	
Diversity/Variety of Choice		NEPA 101(b)(4)
Resource commitments		
Energy Requirements	1502.16(e)	NEPA—102(2)(C)(v)
Natural/Depletable Res.	1502.16(f)	NEPA—102(2)(C)(v), 101(b)(5,6)

For example, there are many toxic and hazardous emissions for which EPA has provided no specific exposure or dose limitation in its National Emission Standards for Hazardous Air Pollutants (40 CFR 61). The EPA has also passed regulatory-based maximum concentration levels (MCLs) for only certain chemical contaminants in water, but it also publishes non-enforceable risk-based benchmarks for many other chemicals, as well as risk-based benchmarks for chemicals having MCLs that are more restrictive than the MCLs. The benchmarks are based on information in peer-reviewed

Chapter eight: Assessing significance

Table 8.4 Applying the receptor test

Step 1: Construct (or expand an agency prescribed) receptor checklist using Table 8.3 as a guide in response to the context of the action.

Step 2: Indicate which receptors are not affected because they are outside the zone of influence of the proposed action, or would otherwise not be affected.

Step 3: Indicate which receptors would be affected at a trivial and clearly nonsignificant level and provide appropriate evidence to this effect.

Step 4: For all other receptors, potentially affected at a nontrivial or potentially significant level, list the issues and the evidence needed to demonstrate nonsignificance or significance.

Table 8.5 Activity-related requirements

Activity	Other federal laws, regs., or reqs.
Generation of Criteria Air Pollutants	Clean Air Act
Generation of Radioactive Air Pollutants	Clean Air Act—NESHAP
Generation of Hazardous/Toxic Air Pollutants	Clean Air Act—NESHAP
Release of Wastewater to Sewage Systems	Clean Water Act
Release of Wastewater to Surface Waters	Clean Water Act
Underground Injection of Waste	40 CFR 146.5
Dredged Materials Discharged to Waters of the U.S.	Sec. 404 Clean Water Act
Underground Storage Tank Installation or Removal	RCRA
Generation/Management of Conventional Waste	RCRA
Generation/Management of Hazardous Waste	RCRA/CERCLA
Generation/Management of Radioactive or Mixed Waste	10 CFR 20
Management and Use of Toxic Substances	40 CFR 700-799
Pesticide Use	FIFRA
PCB Use	40 CFR 761
Asbestos Removal	40 CFR 61 and 763
Disturbance of Pre-Existing Contamination	RCRA, CERCLA
Decontamination/Decommissioning	RCRA Closure Reqmnts.
Noise Propagation	Nuisance Case Law
Generation of Light/Lasers	–
Generation of Ionizing Radiation	–
Generation of Non-Ionizing Radiation/EMF	–
Alteration of Landscape	Zoning and Land Use Laws
Alteration of Water Course	Various Federal Laws

Table 8.6 Applying the activity test

Step 1: Refer to that portion of the agency's NEPA implementation procedures that specify which activities are categorically excluded, which normally require EAs, and which normally require EISs.

Step 2: Construct an activities checklist using Table 8.5 as a guide in conformity with agency practice (such as in Step 1, on prescribed checklists and additional relevant activities meriting consideration).

Step 3: Indicate which activities have no effects because they do not occur in the zone of influence of the proposed action, or other reasons.

Step 4: Indicate which activities occur at a trivial and clearly nonsignificant level, and provide some evidence to this effect.

Step 5: For all other activities that occur at a nontrivial or potentially significant level, correlate with affected receptors and state the evidence needed to demonstrate nonsignificance or significance.

Table 8.7 Applying the regulatory compliance test

Step 1: Determine regulatory compliance requirements using Tables 8.3 and 8.5 as an initial guide for applicable receptors and activities per the previous two tests. Expand the list of applicable compliance requirements to include any appropriate federal, state, and local regulations.

Step 2: Determine whether existing compliance programs (such as permits, licenses, monitoring, and control programs) would also cover the proposal, and cite evidence (such as permit numbers, conditions, etc.).

Step 3: For all receptors not currently covered by compliance programs (i.e., those not cited in steps 1 or 2), determine whether the proposal threatens a violation in principle. This step might require quantifying effects to demonstrate absence of "threat."

---

scientific literature but have not been formally established into regulations in accordance with procedures in the Administrative Procedures Act. In these cases, an agency complying with NEPA needs to go beyond the regulation in assessing significance, and consider the literature on recommended dose and exposure limits such as those published by the EPA and the American Council of Governmental Industrial Hygienists for the workplace (Threshold Limit Values for Chemical Substances and Physical Agents and Biological Exposure Indices, periodically updated):

> The agency may arguably be unable or unwilling to take the necessary steps to be in compliance with all applicable laws, regulations, and requirements.

For example, the agency might have inadequate resources and a poor track record in enforcing compliance.

Risk/Uncertainty Test—The following CEQ significance factor lends itself to objective analysis within certain limitations (§1508.27[b][5]):

> The degree to which the possible effects on the human environment are highly uncertain or involve unique or unknown risks.

This significance factor typically involves two types of cases:

Case 1: This significance factor often involves projects such as hazardous or radioactive treatment facilities in which various accidents can produce consequences ranging up to the catastrophic. Probabilistic risk analysis constitutes a body of available methodology to characterize the risk with some degree of scientific integrity. A procedure based on probabilistic risk analysis is recommended for assessing potential significance in these cases. The reader is referred to Section 7.2 in Chapter 7, which discusses accident analyses.

Case 2: A somewhat less tractable analysis involves cumulative actions whose effects must be considered together with other actions over which the agency has no direct control. In addition, when an action becomes a precedent for other actions, the eventual results can be difficult to predict and defend. Accordingly, both situations can be characterized as highly uncertain. (A discussion of "cumulative effects" and "precedent" is presented shortly.)

Tables 8.8, 8.9, and 8.10 apply to Case 1. Table 8.8 is a modified version of a recommended severity-probability ranking that was used by the U.S. Department of Defense.[3] Table 8.9 is a hazard-frequency scale that has been used by the Department of Energy (DOE).[4] Table 8.10 combines the results of ranking magnitude and frequency (Tables 8.8 and 8.9) into a recommended significance determination. Table 8.10 has been modified slightly by the author.

Recommended practice (for Case 1) for applying the Risk/Uncertainty Test is shown in Table 8.11.

In applying this test, it is important to base a determination of significance on whether risk is increased or decreased compared to an initial or baseline condition. For example, a new hazardous waste storage facility may of itself constitute a severe risk to the environment associated with possible explosions or fires. However, if the proposed action is to move such waste from storage conditions of higher risk to lower risk, the action might arguably be considered nonsignificant even if Table 8.10 indicates otherwise.

The issue of appropriate evidence, including database, methodology, precision, and quality control for characterizing the expected magnitude and frequency of events will vary with the environmental context. As a first-order approach, use of the tables can be based on professional judgment and rule of reason. However, there are circumstances in which regulations

Table 8.8 Severity scale for risk analysis

Severity	Scale	Consequences (human/environmental)
Catastrophic	IV	Human: Loss of ten or more lives and/or large-scale and severe human injury or illness.
		Environmental: Large-scale damage involving destruction of ecosystems, infrastructure, or property with long-term effects, and/or major loss of human life.
Critical	III	Human: Loss of fewer than ten lives and/or small-scale severe human injury or illness.
		Environmental: Moderate (medium-scale and/or long-term duration) damage to ecosystems, infrastructure, and property.
Subcritical	II	Human: Minor human injury or illness.
		Environmental: Minor (small-scale and short-term) damage to ecosystems, infrastructure, or property.
Negligible	I	Human: No reportable human injury or illness.
		Environmental: Negligible or no damage to ecosystems, infrastructure, or property.

Table 8.9 Frequency scale for risk analysis

Category	Level	Description	Frequency (f)
Frequent	A	At least once per year	$f > 10^0$
Likely	B	Once in 1 to 10 years	$10^0 > f > 10^{-1}$
Occasional	C	Once in 10 to 100 years	$10^{-1} > f > 10^{-2}$
Unlikely	D	Once in 100 to 1,000 years	$10^{-2} > f > 10^{-3}$
Remote	E	Once in 1,000 to 1,000,000 years	$10^{-3} > f > 10^{-6}$
Incredible	F	Less than once in 1,000,000 years	$10^{-6} > f$

or an agency might require a formal probability and risk assessment such as DOE's Safety Analysis Reports. Recommended practice is to base the significance determination on such formal analysis when it is available.

When a formal risk analysis is not required by the agency or regulations, the NEPA practitioner must determine whether it is needed in the analysis, and what methodology is to be used. The potential for controversy should be a factor in making these decisions.

The Cumulative Test—The issue of cumulative impacts is one of the most difficult in NEPA practice. Accordingly, application of this test is more complex than the others and merits somewhat more consideration.

Regulatory guidance: The regulations specify consideration of the following cumulative factor (§1508.27[b][7]):

Table 8.10 Recommended significance assignments for severity-frequency determinations

	Frequency					
Severity	A (frequent)	B (likely)	C (occasionally)	D (unlikely)	E (remote)	F (incredible)
IV (Catastrophic)	S	S	S	S	M	M
III (Critical)	S	S	S	M	N	N
II (Subcritical)	S	S	M	N	N	N
I (Negligible)	N	N	N	N	N	N

Note: S = usually significant; M = marginally significant or nonsignificant, depending on context; N = usually nonsignificant.

Table 8.11 Applying the risk/uncertainty test

Step 1: Determine the expected severity of the impacts that could occur for each risk-prone situation associated with an action (i.e., accidents and natural hazards) by referring to Table 8.8.

Step 2: Determine the expected frequency of the impacts that could occur for each risk-prone situation associated with an action (i.e., accidents and natural hazards) by referring to Table 8.9.

Step 3: Determine potential significance as a function of severity and frequency together by referring to Table 8.10.

> Whether the action is related to other actions with individually insignificant but cumulatively significant impacts. Significance exists if it is reasonable to anticipate a cumulatively significant impact on the environment. Significance cannot be avoided by terming an action temporary or by breaking it down into small component parts.

The regulations define a cumulative impact as (§1508.7, emphasis added):

> "Cumulative impact" is the impact on the environment that results from the incremental impact of the action when added to other past, present, and reasonably foreseeable future actions, regardless of which agency (federal or non-federal) or person undertakes such other actions. Cumulative impacts can result from individually minor but collectively significant actions taking place over a period of time.

The language of the regulation thus defines a three-step logic in determining significance of cumulative impacts:

- Create a baseline defined by the environmental impacts of past, present, and reasonably foreseeable future actions independent of the proposed action.
- Determine the incremental impact of the proposed action and superimpose (i.e., "add") it to the baseline.
- Apply the same other six significance tests described in this chapter to the evaluation of cumulative impacts. As described earlier, these are applied in a similar fashion to the way they would be used in assessing the significance of any individual impact.

Tools for assessing cumulative impacts—The common features of analytic tools available for the cumulative test are:

- They characterize a current baseline situation representing the effects of past and present actions.
- They provide a theory for predicting reasonably foreseeable future actions that perturb the current baseline condition.

Some of the principal tools for cumulative impact analysis include:

- Econometric forecasting models for predicting levels of economic activities that would cause environmental impacts.
- Geographic analysis methods that predict land use and related effects.
- Ecosystem models that predict the future state of interdependent living systems.

These tools are often not used to their full potential because:

- Many NEPA practitioners are not familiar with the tools, or are untrained in the associated disciplines of econometrics, economic geography, or ecology.
- Use of these tools may be too data intensive, time consuming, and expensive to apply under many circumstances, particularly for EAs.
- The tools require skilled, experienced specialists who may not be locally available, which further adds to the expense.
- Application of the tools can become controversial in that they are based on assumptions that can be challenged, and their predictive capability is often difficult to demonstrate.

Tools should only be used when they can provide information relevant to significance assessment, and never be used routinely. The time and effort needed to apply a tool should be weighed against the potential value of information gained. Nevertheless, such tools have a place in environmental planning, and can be valuable if used appropriately.

Applying the Cumulative Test—If a finding of nonsignificance with respect to the cumulative test is challenged upon publication of a FONSI, the issues of risk/uncertainty, precedence, and controversy might also be raised.

In applying the Cumulative Test, the recommended practice is outlined in Table 8.12. In applying this test:

- Carefully characterize the environmental issues important for cumulative analysis. Use professional judgment and common sense (i.e., rule of reason) to qualitatively characterize the cumulative effects, or to approximate them using a quantitative method such as one of the tools described above.

Table 8.12 Applying the cumulative test

Step 1: Create a baseline defined by environmental impacts of past and present actions. This baseline is used for comparing the effects of the reasonably foreseeable future actions (including the proposed action).

  a. Define the current environmental baseline. This represents the effects of past and present actions.

  b. Identify all the reasonably foreseeable future actions. Define the impact of reasonably foreseeable future actions, including the proposed action.

  c. Superimpose (i.e., add) the impact of reasonably foreseeable future actions (including the proposed action) on the environmental baseline defined in 1a.

Step 2: Determine whether the impact of the reasonably foreseeable future action, including the proposed action, exceeds the threshold of significance. Apply the other six tests of significance as they would be used in assessing individual impact.

Step 3: Determine whether the incremental impact of the proposed action exceeds the threshold of significance for a given incremental action. The incremental threshold is a measure of the proposed action's contribution relative to the total cumulative effect (past, present, proposed, and reasonably foreseeable actions). By measuring the incremental impact relative to the total cumulative impact, decision makers gain important insight that can enhance project planning and development of mitigation measures. Apply the other six significance tests in a fashion similar to the way they would be used in assessing an individual impact. Note that Step 3 may not be necessary if the significance tests in Step 2 reveal that the total cumulative effect is not significant.

- Obtain assistance from an appropriate expert on the problem area (e.g., economics, ecology, geography, etc.) in selecting and applying an analytic method.

If certain cumulative effects are sufficient to cause a significant impact, and if an EIS is subsequently prepared, expert assistance may continue to be needed in formulating an analytic methodology as part of scoping, and in implementing it during the analysis.

Precedence Test—The CEQ regulation specifies the following significance factor (§1508.27[b][6]):

> The degree to which the action may establish a precedent for future actions with significant effects or represents a decision in principle about a future consideration.

The CEQ has provided no further guidance on how to interpret and implement this significance factor. The first part of this factor is similar to the Cumulative Test in that it requires that the proposal be considered together with (reasonably foreseeable) future actions. Accordingly, appropriate practice for this part of the significance factor is similar to that described for cumulative actions. However, consideration can overlap with the risk/uncertainty factor, given the problems inherent in predicting the future.

The second part of this factor (i.e., decision in principle) is somewhat unique. Appropriate practice can be inferred from court precedent. Fogelman reports:[5]

> This type of effect may occur when construction of a facility—such as a port—ensures that an area will continue to be developed in lieu of other areas. Once the plans are initiated and begun, it is probable that decision makers will order the project continued.

For the practitioner, the "rule of reason" can be used in assessing the "decision in principle." If a finding of insignificance with respect to the Precedence Test is challenged upon publication of the FONSI, the issues of risk/uncertainty, cumulative effects, and controversy might also be raised.

Recommended practice for applying the Cumulative Test is outlined in Table 8.13.

Table 8.13 Applying the cumulative test

Step 1: With respect to precedent for future actions factor, follow the same practice recommended for cumulative impacts.

Step 2: With respect to a decision in principle about a future consideration, use common sense based on an understanding of the action's context.

The Controversy Test—The regulations specify the following factor be considered in assessing significance (§1508.2 [b][4]):

> The degree to which the effects on the quality of the human environment are likely to be highly controversial.

No further guidance is offered by the CEQ in regulation or other published guidance. However, some clear guidance on this issue is provided in the case of *Hanly v Kleindienst (II)*:[6]

> The term "controversial" apparently refers to cases where a substantial dispute exists as to the size, nature, or effect of the major federal action rather than to the existence of opposition to a use, the effect of which is relatively undisputed. ... The suggestion that "controversial" must be equated with neighborhood opposition has also been rejected by others [i.e., other cases].

Thus opposition to the proposed action *per se* does not constitute evidence of "controversy" in the NEPA sense of the term, an extremely important interpretation.

Table 8.14 depicts the recommended practice for applying the Controversy Test.

Table 8.14 Applying the controversy test

Step 1: Determine which issues have the potential to generate controversy based on issues considered in the previous tests.

Step 2: Assess the potential for controversy by knowing the affected stakeholders and potential for project opposition. This assessment should also be based on available public response to notices or news about the proposed action and any history of past opposition on similar issues.

Step 3: As necessary, obtain the assistance of legal counsel to determine whether potential opposition meets controversy criteria per *Hanly v Kleindienst*.

8.3.3 Additional significance considerations in practice

In carrying out the above tests, the focus was on the language in the regulatory definition of significance per §1508.27[a]. Other factors might also need to be considered. One of these factors, the context, is described in the following section.

Context—The regulations require that context and intensity be considered together. It provides intensity but little specific guidance on how to consider context. The language of the regulations does not imply a test of context *per se*, but merely requires that context be considered in applying the various tests defined under intensity.

Context is in part established by the environment in which the action is taking place and can therefore be represented by an environmental receptor checklist as in Table 8.3. It is also established in part by certain features of the action that can also be identified in checklists such as Table 8.5. Thus use of checklists provides evidence that context is considered in part. How much further should the practitioner go in explicitly documenting consideration given to context? This becomes a matter of professional judgment in concert with the "rule of reason," and taking into account any special agency or public concerns likely to be raised during the NEPA process.

Although CEQ provides little guidance on consideration of context in determining significance, the CEQ regulations do note that "Significance varies with the setting of the proposed action. For instance, in the case of a site-specific action, significance would usually depend upon the effects in the locale rather than in the world as a whole." Locally significant impacts that may not seem important in the context of the "big picture of things" can seem quite significant in the context of those who reside close to the site of a project; it is the interests of these local residents that must be considered in NEPA. Theoretically, the apparent significance of almost any impacts can be argued away by spatially "zooming out" from a perspective of locality, region, state, the United States as a whole, or—for really big impacts—the world as a whole. In making the above statement, CEQ is clearly indicating that such an approach is usually inappropriate. By the same token, however, the significance of almost any impacts can be overstated by focusing too narrow on too localized a perspective. Hence there is a need for NEPA preparers to use professional judgment and the rule of reason when determining significance and to carefully document their assumptions and thought processes.

8.3.4 Evidence of significance

Evidence of significance is specifically required in an EA to support a FONSI (§1508.9[a][1], §1500.2[b]). Table 8.15 depicts examples of forms of evidence that have been used in support of FONSIs (listed in approximate order of

Table 8.15 Forms of evidence that have been used in support of FONSIs

- Evidence that the document preparer reviewed items on a checklist.
- Accessible databases and literature pertaining to the affected environment and to individual environmental receptors.
- Official certification or representation by federal, state, or local officials as to whether a given receptor would likely be affected by the action.
- Regulatory, scientific, and professional practice literature pertaining to the effects of certain disturbances (e.g., radiological, chemical) on human health, and on other environmental receptors.
- Computation and modeling to test whether a given impact exceeds a threshold (defined by law, regulation, requirement, or recommended practice) or to characterize intensity of impact.
- Original data collected in the field or laboratory as direct evidence in support of significance or as input to computation and modeling.

rigor, time, and expense). One or more, but not all, of these forms of evidence is necessary. Factors influencing choice of the best form(s) include the character and severity of the impact under consideration, professional judgment of the preparing agency, public controversy, and practicality of use.

The particular choice of evidence for each potential impact considered is at the discretion of the agency, as CEQ has not provided any further guidance other than §1500.1(b), in which the agency is admonished to concentrate on the issues that are truly significant to the action in question rather than amassing needless detail.

8.4 Significant beneficial impacts

As noted in previous chapters, not all environmental impacts are adverse, although adverse environmental impacts were the impetus for enactment of NEPA and are almost always the focus of EISs and EAs. Because NEPA requires consideration of all and not just adverse environmental impacts, the principles of significance determination apply to beneficial as well as adverse impacts. The following section focuses on whether an EIS rather than EA is required if the only "significant" environmental impacts are beneficial ones. Much of the information presented in this section is based on a paper by Swartz.[7]

According to the NEPA Regulations:

> [E]ffects may also include those resulting from actions which may have both beneficial and detrimental effects, even if on balance the agency believes that the effect will be beneficial (40 CFR § 1508.8).

The regulations go on to state that:

> [A] significant effect may exist even if the Federal agency believes that on balance the effect will be beneficial (40 CFR § 1508.27[b][1]).

8.4.1 Hiram Clarke Civic Club v. Lynn

A number of courts have examined whether the existence of beneficial environmental impacts is sufficient to trigger an EIS, with differing results. The earliest case involved a challenge to a proposed low and moderate income apartment project in Houston, Texas.[8] The court concluded that the Department of Housing and Urban Development (HUD) was not required to file an EIS covering a proposed apartment project. The court went on to address the plaintiffs' claim that HUD's determination of "significance" improperly focused only on adverse environmental impacts, contrary to the CEQ guidelines. The court stated:

> [Plaintiffs] argue that NEPA requires that an agency file an environmental impact statement if any significant environmental effects, whether adverse or beneficial, are forecast. Thus, they argue, by considering only adverse effects HUD in effect did but one-half the proper investigation. We think this contention raises serious questions about the adequacy of the investigatory basis underlying the HUD decision not to file an environmental impact statement.

Without elaborating, the court expressed its view that

> [A] close reading of Section 102(2)(C) in its entirety discloses that Congress was not only concerned with just adverse effects but with all potential environmental effects that affect the quality of the human environment.

Despite this, the court agreed that the project in question was not a major federal action significantly affecting the quality of the human environment.[1]

8.4.2 Douglas County v. Babbitt

More recently, a court examined whether the Secretary of the Interior's decision under the Endangered Species Act (ESA) to designate critical habitat for a threatened or endangered species was subject to NEPA.

Chapter eight: Assessing significance 199

The court also concluded that "NEPA procedures do not apply to federal actions that do nothing to alter the natural physical environment." In the court's words, an "... EIS is unnecessary when the action at issue does not alter the natural, untouched physical environment at all."[9]

On appeal, the U.S. Court of Appeals for the 10th Circuit came to a much different conclusion. The 10th Circuit of Appeals held that there were "actual impact flows from the critical habitat designation," and that compliance with NEPA will further the goals of ESA.[10] The court reiterated the plaintiffs' claim that the proposed designation "will prevent continued governmental flood control efforts, thereby significantly affecting nearby farms and ranches, other privately owned land, local economies and public roadways and bridges." The court characterized these impacts as "immediate and the consequences could be disastrous." It went on to state:

> While the protection of species through preservation of habitat may be an environmentally beneficial goal, Secretarial action under ESA is not inevitably beneficial or immune to improvement by compliance with NEPA procedure. ... The short- and long-term effects of the proposed governmental action ... are often unknown or, more importantly, initially thought to be beneficial, but after closer analysis determined to be environmentally harmful.

In another case, the Farmers Home Administration had prepared an environmental assessment (EA) for the funding of a water impoundment and treatment project. The agency concluded that the project would have no significant environmental impacts. However, the agency also concluded that "[t]he project will have a positive impact on the living environment of the residents of the area" because they would be "provided with a dependable, sanitary water supply." The plaintiffs sued, claiming that the existence of "significant" beneficial impacts required the preparation of an EIS.[11] On appeal, the U.S. Court of Appeals for the 6th Circuit held that if an agency concludes, based on an EA, that the project will have no significant adverse environmental consequences, an EIS is not required. The court reasoned that:

- One of the purposes of NEPA is to "promote efforts which will stimulate the health and welfare of man." Since the health and welfare of the residents of Tracy City will not be "stimulated" by the delays and costs associated with the preparation of an EIS "that would not even arguably be required were it not for the project's positive impact."

- The NEPA regulations directs agencies to make the NEPA process more useful to decision makers and the public, to reduce paperwork and the accumulation of extraneous background data, and to emphasize real environmental issues and alternatives (40 CFR § 1500.2(b)).

It is important to note, however, that the court did differentiate between projects where the only "significant" impacts were beneficial ones and projects where there were "significant" beneficial and adverse impacts, but that "on balance" the impacts were beneficial.

8.4.3 Conclusion

Based on these cases, it is not completely obvious whether the requirement to prepare an EIS can be triggered solely by significant beneficial impacts. However, an argument based on the following arguments could be made that significant beneficial impacts can be sufficient to trigger the requirement to prepare an EIS:

- If a project is "large" enough to result in significant beneficial impacts, could it not also produce less obvious or even "hidden" significant adverse impacts? In some cases, an EA may be insufficient to "flesh" out such hidden effects, and therefore an EIS should be prepared.
- A proposal action with significant beneficial impacts (and no significant adverse impacts) could be made even better with the in-depth analysis demanded in an EIS, where alternatives are rigorously investigated and compared.
- Perhaps most importantly, what is perceived as a beneficial impact by some may be perceived as adverse by others. For example, wildlife conservations may be delighted by a proposal by an agency to establish wetlands in an urban area, but some nearby residents may be concerned about possible health impacts caused by mosquitoes attracted to the wetland. The rigorous analysis and enhanced scrutiny resulting from an EIS may lead to mitigation measures such as water aeration or introduction of insectivorous fish that allow the wetland to proceed without increasing the risk for mosquito-borne diseases. A briefer analysis typical of an EA might overlook such creative solutions.

As stated elsewhere in this book, enforcement of NEPA is the responsibility of the agencies themselves, and much of what drives enforcement is the potential for challenges. Logically, adverse impacts drive litigation more than beneficial impacts. Bad news tends to attract more attention than good news. If there are "hidden" adverse impacts accompanying beneficial impacts, the parties adversely affected tend to be increasingly at risk of a legal challenge, as the potential for significance increases. The parties benefiting from the proposal are unlikely to complain. A FONSI that truly covers only beneficial impacts is unlikely to draw the ire of litigation; one that masks adverse impacts under beneficial impacts may attract litigation from parties who recognize that their interests are adversely affected rather than benefited. Agencies must carefully ensure that their FONSIs do not mask significant adverse impacts or eventually they can be expected to face litigation.

References

1. March, F., National Association of Environmental Professionals, 21st Annual Conference Proceedings, *Determining the Significance of Proposed Actions*, pp. 421–36, 1996.
2. Eccleston, C.H., *The NEPA Planning Process: A Comprehensive Guide with Emphasis on Efficiency*, John Wiley & Sons Inc., New York, 1999.
3. U.S. Department of Defense, DOD's MIL-STD-882B.
4. U.S. Department of Energy, Order 5481.1B.
5. Fogelman, V., *Guide to the National Policy Act: Interpretations, Applications and Compliance*, Quorum, 1990.
6. *Hanly v Kleindienst (II)*, 471 F.2nd 823 (2nd Cir. 1972), cert. denied, 412 U.S. 908 (1973).
7. Swartz, L. *Triggering the Environmental Impact Statement Requirement Under the National Environmental Policy Act: Do Beneficial Impacts Count?* Paper presented at the 2001 conference of the National Association of Environmental Professions.
8. *Hiram Clarke Civic Club v. Lynn*, 476 F.2d 421 (5th Cir. 1973).
9. *Douglas County v. Babbitt*, 48 F.3d 1495 (9th Cir. 1995), cert. denied, 64 U.S.L.W. 3167 (January 8, 1996).
10. *Catron County Board of Commissioners v. U.S. Fish and Wildlife Service*, No. 94-2280 (10th Cir. February 2, 1996).
11. *Friends of Fiery Gizzard v. Farmers Home Administration*, 61 F.3d 501 (6th Cir. 1995).

chapter nine

Finding of no significant impacts

> Substitute "damn" every time you're inclined to write "very"; your editor will delete it and the writing will be just as it should be.
>
> **—Mark Twain**

9.1 Introduction

Once an environmental assessment (EA) has been completed, it is read by the decision maker. If the decision maker concludes that the action could not potentially result in any significant environmental impact, this determination is recorded in a finding of no significant impact (FONSI). A FONSI cannot be issued if even one environmental impact is potentially significant.

The FONSI is a decision (i.e., determination) document supported by the analysis in an EA; it may be loosely thought of as analogous to a record of decision (ROD) that is supported by the analysis in an environmental impact statement (EIS). This chapter presents the reader with direction for preparing a rigorous and defensible FONSI.

The analogy between a FONSI and a ROD is not perfect. A FONSI documents a determination that the proposal does not involve a significant environmental impact, and hence the agency's decision to proceed with an action without preparing an EIS. A ROD documents a decision to proceed with one of the alternatives analyzed in an EIS. A FONSI must demonstrate, relying on information in the EA, that the proposal would not potentially result in any significant environmental impacts. A ROD does not have to prove that the selected alternative is not capable of significant environmental impacts; it need only disclose the potential impacts, compare the impacts of reasonable alternatives, and show that the agency has taken a "hard look" at the potential environmental impacts. A ROD need not even demonstrate that the selected alternative has the least significant environmental impacts among the alternatives; it need only document how alternatives were sufficiently considered. Lastly, an EA does not necessarily lead to a FONSI; it could instead lead to a decision to prepare an EIS (or not proceed with the proposal). An EIS generally leads to a ROD, unless the agency decides not to implement the proposal.

Once a FONSI has been issued, circumstances might arise where an agency desires to make a change to the proposed action described in the EA. The situation is somewhat analogous to that of an agency proposing a change to an action after a ROD has been issued. In both situations, the agency may have to supplement its NEPA document (EA or EIS) with additional analysis and issue a new decision document (FONSI or ROD). If the new decision document is a FONSI, it will have to show that the modified proposal still could not potentially result in potentially significant environmental impacts. A decision making tool (termed the Smithsonian Solution) for determining when a change to a proposed action requires additional analysis is presented in Section 8.7.3 of Eccleston's companion text, *NPEA and Environmental Planning*.[1] The Smithsonian Solution is applicable to changes to a proposal covered by an EA and FONSI as well as changes to a proposal covered by an EIS and ROD.

9.2 Reaching a determination of nonsignificance

The purpose of a FONSI is to publicly document a decision maker's determination that, based on review of the EA, the proposal will not result in a significant impact, and therefore preparation of an EIS is not required (§1501.4[e]). The decision maker must assess the environmental facts impartially, avoiding even the appearance of being biased or prejudicial. The FONSI should therefore be prepared after the EA is fully written and all relevant decision makers in the agency have had a chance to read the EA and any associated supporting technical studies; agencies preparing an EA should remain open to the possibility of preparing an EIS throughout the entire process of completing requisite analyses and writing the EA.

The concept of "significance" lies at the heart of an EA and FONSI; in contrast, disclosing potentially significant impacts and comparing alternatives are the primary purposes of an EIS and ROD. While most agencies avoid even the mention of significance in their EAs, the FONSI is where significance can and must be front and center. The definition of significance and the factors used in reaching a significance determination are described in Chapter 4 of this text. Chapter 8 of this text provides the reader with a systematic framework and methodology for assessing the significance of an impact.

In some instances, proposed actions evaluated in an EA are determined to have associated beneficial (sometimes termed "positive") environmental impacts. Like adverse (or "negative") impacts, beneficial impacts can be deemed significant if they clearly and substantially cause environmental conditions substantially different than would have happened had the action not been implemented. No specific guidance has been developed for documenting beneficial impacts in FONSIs, but clearly FONSIs should address all impacts, adverse or beneficial. The CEQ does not specifically indicate whether significant beneficial environmental impacts preclude the issuance of a FONSI, but in practice the focus is on potentially significant adverse impacts that preclude FONSI issuance. As NEPA's enforcement lies largely in the courts and public participation, implementation of an action with only beneficial environmental impacts is unlikely to trigger public outrage, regardless of whether supported by an EA and FONSI or an EIS and ROD. However, actions with significant beneficial impacts to some resources and significant adverse impacts to other resources clearly cannot be supported by an EA and FONSI. Significant beneficial impacts do not cancel out significant adverse environmental impacts for purposes of qualifying an action for a FONSI unless they directly and completely mitigate the adverse impacts. For example, an action with significant beneficial socioeconomic impacts but also with significant adverse ecological impacts does not qualify for a FONSI.

It is considered good professional practice to also include significance determinations in the FONSI for each reasonable alternative evaluated in detail in the EAs. But while FONSIs must demonstrate that the proposed action would not result in significant environmental impacts, they do not have to conclude nonsignificance for the alternatives. In fact, it is common for FONSIs to conclude that one or more alternatives evaluated in the EA could cause significant environmental impacts. Such a conclusion actually bolsters the argument for proceeding with the proposed action. Of course, if the agency were to actually seek to change from the proposed action to an alternative presented in the EA as having significant impacts, it would have to prepare an EIS rather than issue a FONSI.

9.2.1 Principles governing sound decision making

A FONSI, like a ROD, is a finding of fact. As such, a FONSI should be a citation of specific facts that the decision maker finds to be true and leads to the conclusion that the proposed action would not result in a significant environmental impact. Principles governing the decision maker's conduct in reaching a FONSI are presented in Table 9.1.

9.2.2 Criteria that must be met in reaching a decision to not prepare an EIS

As indicated in Table 9.2, the courts appear to have established four criteria that must be met in the process of reaching a decision to not prepare an EIS:[2]

The decision maker must carefully weigh the evidence presented in the EA before concluding that a FONSI is appropriate. The FONSI should be written to clearly demonstrate to the reader that the responsible

Table 9.1 Principles governing the decision maker's assessment for a FONSI

The agency bears the *burden of proof* in demonstrating that no significant impacts would result; otherwise the agency must abandon the proposal, modify the proposal or proposed mitigation that avoids or offsets significant environmental impacts, or proceed to prepare an EIS.

The decision maker's role is **not** to be a *committee of compassion* for the project proponent.

The final decision **must** be based on the *assessment of facts*.

Information (particularly with respect to an applicant) is not the same as facts.

Opinions or best professional judgment without some factual basis are *without merit*.

Table 9.2 Four criteria established by the court that agencies must meet in reaching a decision not to prepare an EIS

1. Whether the agency took a "hard look" at the problem.
2. Whether the agency identified the relevant areas of environmental concern.
3. Whether the agency made a convincing case that the impact was nonsignificant.
4. If there is an impact of true significance, has the agency convincingly shown that changes in the proposed action would sufficiently reduce the effect to the point of nonsignificance?

decision maker signing the FONSI thoroughly understands (1) the scope of the action; (2) implications of the proposal; and (3) that no significant environmental impacts would result.

It is worth noting that a FONSI is limited to consideration of the significance of environmental impacts, i.e., those impacts falling in the purview of NEPA. Significance of cost impacts, feasibility impacts, and other impacts not falling in the purview of NEPA usually play key roles in decisions to proceed with most actions but should not drive a decision to issue a FONSI or influence the decision whether or not to proceed with an EIS.

As an example of a poor decision making process, a recent court case illustrated both the importance of preparing an EA that is readable and the need for the decision maker to thoroughly understand the analysis before reaching a final decision. This case involved a state-sponsored action that required federal authorization. The action was subsequently challenged in court. In a deposition, the decision maker was asked to read a highly technical section from the document involving the analysis of an environmental hazard. The decision maker acknowledged that he did not completely understand this analysis. When asked about the significance of certain impacts that had been analyzed, he also acknowledged that he was not sure how many people could be harmed in the event of an accident. The decision maker did not appear to thoroughly understand the analysis for which he had made a decision. Sensing that it was in a weak position, the agency eventually agreed to settle the case out of court.

9.2.3 Agency's administrative record

An agency's administrative record (ADREC) memorializes the consideration of all relevant and reasonable factors that were used in evaluating and reaching a final decision (which for an EA is the decision to issue a FONSI or to prepare an EIS). Overall, the ADREC should demonstrate and document that the agency examined the proposed action and its reasonable alternatives thoroughly as required by law. Records management is important for two reasons:

1. To satisfy legal requirements.
2. To enable assembly of documents in litigation.

The concept of an ADREC comes from the judicial review section of the U.S. Administrative Procedure Act (APA). The APA requires public disclosure on federal rulemaking efforts and other actions that have the effect of rulemaking. This requirement has a stepped process, similar to NEPA, where rules are published first in draft form. After the public has had an opportunity to submit comments on the proposed rule, a final rule is published. The concept of an administrative record comes from the judicial review section of the APA.

The ADREC should consist of relevant and significant documents considered by the decision maker in reaching a final decision. If the document is irrelevant or insignificant, it should not be included in the ADREC. At a minimum, the following types of information should be included in an administrative record:

- Background documents that help explain the context in which the decision was made.
- Documents relied on by the decision maker, or incorporated by reference in documents relied on by the decision maker, whether or not those documents support the final agency decision.
- Comments received during the public review process from other agencies and the public.
- Responses to comments received during the public review process.
- Summaries of meetings with the public to discuss the proposed action.

Table 9.3 Generalized checklist for reaching a determination of nonsignificance

---
Has the agency taken a "hard look" at the problem?

Have all relevant areas of environmental concern been adequately identified?

Has the decision maker impartially assessed the facts with an open mind?

Did the decision maker carefully weigh the evidence and facts presented to him before concluding that a FONSI was appropriate?

Does the decision maker thoroughly understand the scope of the action, its implications on environmental quality, and the fact that no significant impact would occur?

Has the agency made a convincing case that all impacts would be nonsignificant?

If there is a significant impact, has the agency convincingly shown that changes in the proposal or other mitigation actions would sufficiently reduce the effect below the threshold of significance?

Was the final determination made based on an assessment of the facts?

Does the agency's administrative record fully, clearly, and accurately document how the decision maker progressed through the facts to arrive at a final determination?

The text *Environmental Impact Statements* details the requirements for preparing and maintaining a defensible administrative record. Table 9.3 provides a generalized checklist that can be used by agency officials in reaching a determination of nonsignificance.

9.3 Preparing the FONSI

A poorly prepared FONSI is vulnerable to legal attack, more directly so than the supporting EA. As the agency is faced with the burden of proving that no significant impacts would occur, all conclusions presented in the FONSI must be directly tied to the analysis presented in the EA. A FONSI must not present conclusions that are not supported by its EA, and neither should a FONSI present new technical data or analytical results not included in its EA. If a need for new data or analyses is discovered in the process of writing a FONSI, then the requisite information should be added to the EA.

As emphasized in Section 9.1, the FONSI must focus on demonstrating the lack of potential significant impacts from the proposed action. The FONSI must tie its conclusions to technical information in the EA, but it does not need to, and should not, duplicate the contents of the EA. A good FONSI must go beyond discussing the technical background of the proposed action (that is the function of the EA) and use that information to demonstrate the lack of potentially significant environmental impacts.

FONSIs that merely discuss technical information that properly belongs in an EA without proceeding to the next step of significance determination fail their purpose.

9.3.1 Documentation requirements

At a minimum, the FONSI must contain the items indicated in Table 9.4 (§1508.13). With respect to appending a copy of the EA, the FONSI may either (1) include, (2) summarize, or (3) incorporate it by reference.[3] Ideally, the best FONSIs incorporate the EA by reference and use technical information from the EA to explain the significance determinations. Table 9.5 describes four basic attributes that every FONSI should address.

In addition to the regulatory requirements described above, every FONSI should:

- Explicitly state that the proposed action would not result in any significant environmental impacts.
- Specifically cite or summarize evidence indicating that no significant environmental impacts would occur. This evidence should discuss both the (1) intensity and (2) context of the impacts that were considered.
- Indicate which factors were weighted most heavily in making the determination.

FONSIs are usually bound within the covers of their EA, either in the front or the back. They may be prepared as entirely separate documents, although no FONSI can be properly understood without the concurrent availability of its EA. Nevertheless, FONSIs are often read by people

Table 9.4 Regulatory requirements for the FONSI

1. A brief explanation of the reasons why the action would not have a significant effect on the human environment. If the EA is included, this discussion need not repeat discussion in the assessment but can incorporate it by reference.
2. Include the EA or a summary of it.
3. Note any other key environmental documents related to the scope of the proposed action.

Table 9.5 Where, who, when, and why

The FONSI should include information clearly indicating the scope of the action and *where* it is to take place. It should also indicate *who* has proposed the action, *why* the action was proposed, and *when* it is scheduled to be carried out.

lacking the time or interest to read the EA and should therefore open with a brief summary of the purpose and need and proposed action and alternatives. The FONSI text should be written following the style guidance for a stand-alone document.

FONSIs must do more than merely state that an action would not result in significant environmental impacts; they must substantiate that conclusion. They must draw upon information and analytical results presented in the EA to support the stated conclusion. FONSIs should not present new analytical data or results not included in the EA; if the EA is inadequate, it should be modified accordingly before issuing the FONSI.

Conciseness and brevity are even more important in writing a FONSI than in writing an EA. Many more people normally read the FONSI than read the EA for an action. The sole purpose of a FONSI is to document a significance determination. A FONSI may summarize but should not repeat technical information presented in the EA. A FONSI must focus on the significance determination and should not read like a cover page or executive summary to the EA. Even more than an EA, FONSIs should avoid technical jargon and technical terms that are not understood by the general public.

Although CEQ has not issued guidance regarding page length for FONSIs, FONSIs for most simple proposals addressed in short EAs should be only 1 or 2 pages. FONSIs for some complex projects addressed in longer EAs might have to be longer, perhaps up to 4 or 5 pages. Excessively wordy FONSIs may not only confuse readers but also convey an impression that the agency lacks confidence in its significance determination. As for EAs, the length of FONSIs should be based on a sliding scale relative to the complexity and controversy of the proposed action. As a very general rule of thumb, a FONSI should ideally be less than 10 percent the length of its EA (without considering attachments and appendices) and never be more than 15 percent.

Most FONSIs consist only of text, without tables, graphics, photographs, or other visual aids. However, use of these features in FONSIs is not expressly discouraged by the CEQ or most agencies implementing regulations. FONSIs may, however, refer to graphics or tables in the EA without duplicating them. The EA is usually the only document referenced in a FONSI. References to original source documents are usually (and best) relegated to the EA, although citation of other references is not expressly prohibited. As for most other technical writing, use of bulleted lists and short paragraphs can improve the readability of a FONSI.

Decisions contained in the FONSI must always be made by the responsible agency, even when the EA is prepared by a contractor. Most agencies, when hiring a contractor to write an EA, typically assign the contractor to draft a FONSI as well. That approach is fine as long as the agency remembers that the decisions contained in the FONSI are those of the agency, not the contractor. Agencies must be especially careful when reviewing FONSI text drafted by their contractors. FONSIs must be signed by a responsible official of the agency; the federal decision maker must read the FONSI very carefully to ensure that it accurately reflects the view of the agency. Reputable contractors will prepare the EA impartially, and will usually provide their opinion to the agency regarding significance of impacts. But the significance determination is ultimately the decision of the agency, not the contractor. Effective contractors will always discuss any potentially significant impacts with the agency upfront, upon discovering significant impacts; hence FONSIs written at the conclusion of the EA writing phase should usually not contain surprises.

9.3.2 A checklist for preparing the FONSI

A generalized checklist for assisting an agency in preparing the FONSI is presented in Table 9.6.

9.4 Mitigated FONSIs

Mitigation measures are additional steps, not part of the proposed action, that can be taken to eliminate or reduce potential impacts (§1508.25[b][3]). Actions considered to be standard engineering practice or required under law or regulation are not normally considered mitigation measures.[4] However, the difference between standard engineering practices and bona fide mitigation measures is not sharply delineated; the only important point is identifying what actions will be taken to reduce impacts. Hence best management practices such as the standard erosion control practices required by many states and municipalities may not strictly meet the definition of mitigation measures, but presenting them as mitigation measures is not a serious mistake. The only serious mistake would be failure to acknowledge the practices in the EA/FONSI or subsequent failure to actually implement the practices. The term "impact reduction" is frequently used in referring to measures used for reducing adverse impacts regardless of whether such impacts are significant or not.

Agencies are encouraged to include impact reduction measures in an EA, even if the effects are not considered significant.[5] Such measures may allow an agency to further reduce impacts even though they might already lie below the threshold of significance. Taking those measures may help reduce the potential for significant future cumulative impacts not presently anticipated or predictable based on information currently available. Additional information on mitigation can be found in Eccleston's companion text, *NEPA and Environmental Planning*.[6]

Table 9.6 Generalized checklist for preparing the FONSI

Has the FONSI been specifically prepared and tailored to the action described in the EA?

Does the FONSI clearly demonstrate that the responsible decision maker thoroughly understands the scope and comprehends the environmental implications of the action, and that no significant environmental impacts would result?

Does the FONSI explain the scope of the action, including any connected actions (e.g., who, what, when, where, why, and how)?

Does the FONSI include the EA or a summary of it?

Does the FONSI explicitly state that no significant impacts will result from pursuing the action?

Does the FONSI conclusively demonstrate and explain why the action will not result in any significant environmental impacts (direct, indirect, or cumulative)?

Does the FONSI demonstrate that both the intensity and context were taken into account in reaching a decision of nonsignificance?

Does the FONSI indicate which factors were weighed most heavily in reaching the determination of nonsignificance?

Does the FONSI indicate what reasonable alternatives were evaluated?

Are all conclusions directly tied to the analysis presented in the EA?

Does the FONSI note any other environmental documents related to the scope of the action?

Does the FONSI describe any mitigation measures that will be adopted? Were such measures designed and customized to address the specific impacts? If applicable, is a specific monitoring and implementation plan described to ensure that any mitigation measures are successfully adopted?

Has the EA adequately evaluated the effectiveness of any mitigation measures committed to in the FONSI? Would these measures mitigate any significant impacts to the point of nonsignificance? Are these mitigation measures free from scientific controversy?

Do funding and technical means exist for implementing any mitigation commitments made in the FONSI?

The decision maker should carefully consider any mitigation measures that are to be adopted as such commitments are legally binding. The agency is responsible for ensuring that funds and technical means exist for implementing such measures. If the proposed measures are not technically or financially practicable, then they may not be considered when issuing a FONSI. If the impacts would be significant without such unreasonable mitigation measures, then the action can only proceed with an EIS.

9.4.1 Mitigation and the courts

At one time, CEQ discouraged the use of the mitigated EA/FONSI.[7] Actions that could result in significant environmental impacts require an EIS, even if mitigation measures could be practicably implemented to reduce the impacts to nonsignificance. During the 1980s and 1990s, numerous courts upheld an agency's right to prepare mitigated FONSIs.[8] Today, a majority of appellate courts have accepted the use of mitigated FONSIs. The acceptability of mitigated FONSIs is one factor that has contributed to the longer EAs prepared in recent years.

Mitigation may be deemed insufficient if the agency has no control over how it will be implemented due to reliance on a third party to implement the measures. When mitigation measures will be implemented by a third party, the commitment, while it need not be contractual, must be more than vague statements of good intentions.[9] Mitigation is generally deemed legally sufficient if it reduces the impact to the point of nonsignificance and is offered by an agency with control to legally enforce the measures.[10] To support a FONSI, mitigation measures must be practicable and certain. If doubt remains as to whether the measures can be successfully implemented, the FONSI may not be supportable and an EIS may be necessary.

Examples where mitigation has been upheld—The courts have upheld mitigation in cases such as airport noise suppression, incorporation of water release procedures for maintaining lake levels during spawning season, and replacing lost wetlands.[11] For example, an EA prepared by the Army Corp of Engineers in 1993 for a hydroelectric project was challenged. The court found that the proposed mitigation was sufficient to compensate for the adverse impacts, even though these measures would not completely compensate for the adverse effects.[12]

In another case, a mitigated EA was prepared for transportation of hazardous materials by rail. When challenged, the court concluded that the mitigation measures were sufficient to support a FONSI.[13] The mitigation included such measures as:

1. Notifying surrounding communities of the train schedule.
2. Providing surrounding communities with copies of the emergency response plans.
3. Providing a toll-free telephone number to local emergency response groups.
4. Developing the proposed action in accordance with the Department of Transportation's regulations for transport of hazardous materials.

Examples where mitigation has been rejected—In contrast, the courts are unlikely to accept mitigation measures that (1) have not been adequately investigated or (2) cannot conclusively demonstrate that the impacts would be reduced to the point of nonsignificance. For example, preservation of a canyon from a river diversion project was found to be inadequate because it did not mitigate effects on species located outside the preserved area.[14] As another example, an EA was prepared for construction of a parking lot that involved removal of vegetation and 500-year-old cedar trees that the agency's administrative record characterized as significant. The EA included certain mitigation measures. The court addressed the issue of mitigation by noting that an agency "may reach a FONSI if mitigation measures are proposed that directly address the impacts identified in the EA." In this case, however, the agency's mitigation—removal of a nearby picnic area and its regeneration as forest (that would take more than 500 years to recover)—lacked adequate scientific analysis to constitute sufficient mitigation to support a FONSI.[15]

Wetland mitigation required as part of permits for wetland impacts under the Clean Water Act or state wetland acts is a common component of mitigated FONSIs for construction projects. FONSIs should do more than merely state that mitigation would be developed to the satisfaction of the permitting agency(ies). The EA should describe the proposed mitigation and how that mitigation reduces wetland impacts to nonsignificance, and the FONSI should include a brief summary of the mitigation and its suitability in reducing impacts. Since 2008, the U.S. Army Corps of Engineers, which issues Section 404 permits under the Clean Water Act, has encouraged the purchase of credits in commercially established offsite wetland mitigation "banks" in lieu of the traditional permittee-sponsored project-specific wetland mitigation projects.[16] EAs and FONSIs for projects relying on wetland mitigation banks should identify the specific bank where credits will be purchased, confirm that the credits are available, and indicate how the credited wetland acreage will offset significant wetland impacts.

It must also be noted that meeting wetland mitigation requirements established under Clean Water Act permits may not necessarily constitute adequate mitigation to support a mitigated FONSI. Not all wetlands are under Clean Water Act jurisdiction,[17] nor are all types of wetland impacts regulated under the Clean Water Act. Isolated wetlands lacking a surface connection to navigable waters are not protected under the Clean Water Act. Wetland impacts not caused by dredge and fill activity, such as sedimentation or fugitive dust deposition, are not regulated under the Clean Water Act. But impacts to isolated wetlands and wetland impacts not related to dredge and fill can still be environmentally significant and require mitigation to support a FONSI.

Table 9.7 Six basic criteria that the analysis of mitigation measures must address

1. Proposed mitigation measures must be demonstrably effective. Cursory statements regarding the effectiveness of mitigation measures are generally insufficient; an EA must present sufficient evidence to support mitigation claims.
2. The effectiveness of mitigation measures should be free from scientific controversy. Disagreement or controversy among experts can substantially weaken an agency's ability to prove that no significant impacts will occur.
3. Mitigation measures should address specific environmental issues and concerns, including cumulative impacts. Mitigation measures that are vague or general in nature are normally inadequate. More to the point, mitigation measures must be designed to address specific impacts and issues.
4. Mitigation measures should be fully identified and defined prior to filing out the FONSI. An agency cannot rely on future or to-be-determined mitigation measures because there is no way to adequately assess whether such methods would *effectively* mitigate the environmental impacts.
5. Mitigation measures must effectively reduce impacts to the point of nonsignificance.
6. A specific monitoring or implementation plan should be included to ensure that mitigation measures are effectively carried out.

9.4.2 Criteria for adopting mitigation measures

The courts appear to be imposing six criteria (Table 9.7) in the analysis of mitigated EAs.[18] Before issuing a FONSI relying on mitigation measures, it is recommended that the decision maker ensure that the mitigation measures are consistent with the six criteria.

References

1. Eccleston, C.H., *NEPA and Environmental Planning: Tools, Techniques, and Approaches for Practitioners*, John Wiley & Sons Inc., New York, 2009.
2. *Cabinet Mountains Wilderness/Scotchman's Peak Grizzly Bears v Peterson*, 685 F.2d 678 (DC Cir. 1982).
3. CEQ, Council on Environmental Quality—Forty Most Asked Questions Concerning CEQ's National Environmental Policy Act Regulations (40 CFR 1500–1508), *Federal Register*, Vol. 46, No. 55, 18026–38, March 23, 1981, Question number 37a.
4. CEQ, Public memorandum, *Talking Points on CEQ's Oversight of Agency Compliance with the NEPA Regulations*, 1980.
5. CEQ, Council on Environmental Quality—Forty Most Asked Questions Concerning CEQ's National Environmental Policy Act Regulations (40 CFR 1500–1508), *Federal Register*, Vol. 46, No. 55, 18026–38, March 23, 1981, Question number 39.

6. Eccleston, C.H., *NEPA and Environmental Planning: Tools, Techniques, and Approaches for Practitioners*, John Wiley & Sons Inc., New York, 2009.
7. CEQ, Council on Environmental Quality—Forty Most Asked Questions Concerning CEQ'S National Environmental Policy Act Regulations, 40 CFR 1500–1508, *Federal Register*, Vol. 46, No. 55, 18026–38, March 23, 1981.
8. *Hawksbill Sea Turtle v. FEMA*, 126 F.3d 461 (3rd Cir. 1997); *City of Waltham v. U.S. Postal Service*, 11 F.3d 235 (1st Cir. 1993); *Roanoke River Basin Ass'n v. Hudson*, 940 F.2d 58 (4th Cir. 1991); *Abenaki Nation of Mississquoi v. Hughes*, 805 F. Supp 234 (D.Vt 1992).
9. *Preservation Coalition v. Pierce*, 667 F.2d 851 (9th Cir. 1982).
10. *Louisiana v. Lee*, 758 F.2d 1081 (5th Cir. 1985).
11. *CARE Now, Inc. v. Federal Aviation Administration*, 844 F.2d 1569 (11th Cir. 1988); *Roanoke River Basin Association v. Hudson*, 940 F.2d 58 (4th Cir. 1991); *Abenaki Indian Nation v. Hughes*, 805 F. Supp. 234 (D. Vt. 1992).
12. *Friends of Pavette v. Horseshoe Bend Hydroelectric*, 988 F.2d 989 (9th Cir. 1993).
13. *City of Auburn v. U.S. Government*, 154 F.3d 1025 (9th Cir. 1998).
14. *United States v. 27.09 Acres of Land (II)*, 760 F. Supp. 345 (S.D.N.Y 1991); *Morgan v. Walter*, 728 F. Supp. 1483 (D. Idaho 1989).
15. *Coalition for Canyon Preservation and Wildlands Center for Preventing Roads v. Department of Transportation*, No. CV 98-84-M-DWM, 1999, U.S. District., LEXIS 835 (D. Mont. January 19, 1999).
16. Compensatory Mitigation for Losses of Aquatic Resources, Final Rule, *Federal Register*, Vol. 73, No. 70, April 10, 2008, pp. 19594–19705.
17. Clean Water Act jurisdiction following the U.S. Supreme Court's decision in *Rapanos v. United States & Carabell v. United States*, U.S. Environmental Protection Agency and U.S. Army Corps of Engineers, June 5, 2007, available at http://www.epa.gov/owow/wetlands/pdf/RapanosGuidance6507.pdf.
18. Daniels, S.E. and Kelly, C.M., Deciding between an EA and an EIS may be a question of mitigation, *Western Journal of Applied Forestry*, Volume 5, Number 4, March 1991.

section two

NEPA case law and non-NEPA environmental assessment documents

chapter ten

An overview of NEPA law and litigation

> Why shouldn't truth be stranger than fiction? Fiction, after all, has to make sense.
>
> —Mark Twain

10.1 Principles underlying legal interpretations of NEPA

This chapter outlines some basic legal principles and their application to litigation for bringing suits based on the National Environmental Policy Act (NEPA). These topics are considered in terms of preparing both environmental assessments (EAs) and environmental impact statements (EISs), although the principal focus is on the EA. As stated in previous chapters of this book, lawsuits and the threat of lawsuits are the primary drivers forcing agencies to comply with NEPA. Unlike many other environmental laws in the United States, neither the U.S. Environmental Protection Agency (EPA) nor any other single agency has primary enforcement authority under NEPA. Instead, each federal agency whose actions are covered under NEPA is responsible for self-enforcement. Not all decision makers in all covered federal agencies welcome their responsibilities imposed by NEPA. Often, lawsuits or the threat of lawsuits are the only barrier to skipping, short-cutting, or improperly carrying out the requirements of NEPA.

The EA and finding of no significant impact (FONSI) became established as a shortcut to many of the procedural burdens of preparing EISs for agencies proposing relatively simple actions for which any environmental actions can be demonstrated to not be significant. The option of preparing an EA and FONSI in lieu of an EIS can be very tempting to agencies proposing actions on tight budgets or schedules, even when those actions might result in potentially significant environmental impacts. Conversely, many environmental activists are often suspicious when an agency elects to prepare an EA and FONSI instead of following the more rigorous EIS path. The issue of lawsuits, while applicable to all NEPA practitioners, is therefore of particular interest to the EA practitioner. For a more detailed treatment of NEPA law, the reader is referred to Mandelker's book, *NEPA Law and Litigation*.

10.2 Process for passing laws and regulations

A study of NEPA court cases in 2009 found that an agency usually won where it could demonstrate it had given potential environmental impacts a "hard look."[1] Before addressing the subject of NEPA litigation, a brief explanation of the process followed in promulgating laws and regulations is instructive. U.S. statutes are enacted by Congress as public laws and are generally signed into law by the president. Statutes are written laws enacted by a legislature (as opposed to oral or customary law).

When a congressional act (including but not limited to those involving environmental issues) becomes law, its first official published form is a pamphlet called a "slip law." Slip laws are cited by giving the public law number, the Congress, and the date of its enactment. Thus the Federal Water Pollution Control Act Amendments of 1972 were approved on October 18, 1972, as the 500th law of the 92nd Congress. As cited in the slip law, this act would appear as "Pub. L. 92-500, October 18, 1972."

10.2.1 United States Statutes at Large

The slip laws from each session of Congress are collected by the office of the Federal Register and annually copied verbatim from the original acts in a volume titled *The United States Statutes at Large*. Thus all slip laws approved during a given session of Congress are published in a bound set known as the *United States Statutes at Large*; it contains public and private laws, resolutions, and proclamations.

Once a law is published in the *United States Statutes at Large*, it will then be cited using the *United States Statutes at Large* nomenclature rather than the slip law. Thus the text of the Federal Water Pollution Control Act Amendments of 1972 appears in Volume 86 of the *United States Statutes at Large* (page 816–905). It will subsequently be cited as "86 Stat. 816-905 (1972)."

Unfortunately, the *Statutes at Large* is not a convenient tool for research. It is arranged strictly in chronological order. Thus statutes addressing related topics, including but not limited to environmental topics, may be scattered across many volumes. These statutes often repeal or amend earlier laws, and extensive cross-referencing is required to determine what laws are in effect at a given time. Those attempting to use the *United States Statutes at Large* are faced with an illogically organized mass of legislation for which the applicability, meaning, and operative force cannot be easily determined. For this reason, the *United States Statutes at Large* is not the source most frequently consulted by environmental practitioners and others having need to consult United States laws. This is because another and much more logically organized code that is far easier to use has been devised.

10.2.2 United States Code

Session laws published in the *United States Statutes at Large* have been codified into the *United States Code* or "U.S.C." The *United States Code* is divided into titles, each pertaining to a distinct topic. For example, the Federal Water Pollution Control Act, as amended, is now referred to as the Clean Water Act, and is codified at Title 33 of the *United States Code*, Sections 1251 to 1387. This means that the text of the Clean Water Act is to be found at 33 U.S.C., Sections 1251 to 1387. Many, but by no means all, environmental laws such as NEPA and the Clean Air Act were incorporated into Title 42 of the U.S.C. As a basis for exploring the use of the *United States Statutes at Large* and the *United States Code*, consider the NEPA, which is often cited with its Public Law (P.L.) citation followed by its U.S.C. citation:

> National Environmental Policy Act (1970), P.L. 91-190, codified as amended at 42 U.S.C., Sections 4321-4347.

The *United States Code* was created to make finding relevant and effective statutes simpler by reorganizing them by subject matter, and eliminating expired and amended sections. Thus the U.S.C. is a logical compilation and codification of all current federal laws of the United States. One may think of it as a frequently updated "running list" of all U.S. laws and their contents, organized conveniently by topic. Whenever existing laws are changed or rescinded, *United States Code* is updated so that the content of each law grouped under each topic remains current. While the *United States Code* is the most frequently consulted source for United States law, the Supreme Court has ruled that the *United States Statutes at Large* is the official source for United States laws. In other words, if the text of a law as it appears in the *United States Code* is at variance with the text as it appears in the *United States Statutes at Large*, the latter version governs.

Title 1 of the U.S.C. outlines the general provisions of the *United States Code*. The entire U.S.C. contains 50 titles. In this context, the word "title" is roughly akin to a printed "volume," although many of the larger titles span multiple volumes. These titles may optionally be divided into subtitles, parts, subparts, chapters, and subchapters. All titles have sections represented by a "§." Section numbers cited in the public law do not necessarily follow the same section numbering used in the *United States Code*.

10.2.3 Code of Federal Regulations

Administrative law is a common-sense recognition that government must be administered efficiently. To this end, Congress must delegate some of its authority to the agencies with the scientific or other technical expertise for transacting such laws. Many federal agencies are given rule-making power as well as authority to make and enforce decisions under the rules. Many Congressional laws require certain federal agencies, often the EPA, to issue implementation regulations, which provide detailed requirements for implementing the broad requirements provided in the statute. For example, Congress might pass a statute that simply states there are not to be "excessive" levels of mercury in any significant U.S. water body (but provides no specific concentration limits or other restrictions or measures). Such a statute may designate an entity (such as the EPA) with responsibility for enforcing the statute and for defining scientifically established levels of acceptable mercury concentrations as well as other constraints and requirements, as well as what constitutes a "significant U.S. water body." The EPA's definitions, penalties for violations, and enforcement provisions implementing what Congress intended will be published as a regulation in the Code of Federal Regulations (CFR).

Moreover, congressional legislation can be passed, which gives an entity wide latitude in creating rules (law of bases). For example, Congress could authorize EPA to pass stringent rules that "control harmful pollutants"; the EPA could then establish broad rules (including definitions and enforcement provisions), in the absence of existing specific laws, to control pollutants such as radon, pesticide, or lead emissions. Such rules are published in the CFR. The fact that elected officials must rely on unelected technical experts employed by federal agencies to develop and implement regulations is highly controversial, especially in environmental topics. When political critics complain about "unelected bureaucrats writing laws," they are in fact complaining about the need for elected officials who generally lack relevant technical expertise to rely on unelected experts employed by agencies to write regulations for the laws the officials pass.

Generally, the agency first prepares a draft implementing regulation published as a proposed rule in the Federal Register (FR). The public is invited to review and submit comments on the draft regulation within a certain specified period (usually 90 days). These public comments are considered by the agency, and if deemed worthy, are incorporated into the final regulation, which is then published in the Federal Register. Note the similarity to the process of publishing a draft EIS for public comment prior to issuance of a final EIS and record of decision (ROD).

Just as session laws published in the *United States Statutes at Large* have been codified in the *United States Code*, similarly the administrative laws published in the Federal Register are codified in the Code of Federal

Regulations. Thus following its publication in the Federal Register, such regulations are codified annually in the CFR. The CFR is the codification of the rules and regulations (sometimes called administrative law) published in the Federal Register by the U.S. federal executive departments and agencies. A CFR is referenced by its "Title" and "Part" numbers. For example, the Council of Environmental Quality's NEPA regulations are provided in "Title 40 CFR Parts 1500-1508" or more simply "40 CFR Parts 1500-1508." A section within the CFR is analogous to a paragraph in a text and normally designates a single thought. For example, in 40 CFR Parts 1500.1, the section ".1" specifies the purpose of the regulation. In this book, it would be shortened to a more convenient expression of "(§1500.1)"

The CFR is divided into 50 titles representing broad subject areas. For example, Title 10 contains regulations pertaining to energy, while Title 40 deals with environmental issues. Title 40 is divided into a number of distinct chapters. Each chapter contains all of the regulations that were promulgated by a certain agency. The bulk of Title 40 is devoted to regulations issued by the EPA, which are contained in Chapter 1. The Council on Environmental Quality's NEPA regulations are contained in Title 40, Chapter 5.

10.3 Interpreting NEPA's regulatory requirements

Readers must sometimes exercise a considerable degree of interpretation in complying with NEPA's requirements. An overly narrow view of a regulatory provision can lead to incorrect interpretations. Nor should one interpret a particular provision insulated from the rest of the act or its regulatory requirements. Specifically, the provisions of the act and of these regulations must be read together as a whole in order to comply with the spirit and letter of the law (§1500.3). For example, discerning the proper meaning and applicability of basic concepts such as "federal actions" and "significance" requires professional experience and interpretation, combined with knowledge of case law. Dickerson describes two principles that are also applicable to the interpretation of environmental regulations:[2]

- A literal interpretation is the meaning carried by language when it is read in its dictionary sense, unaffected by a particular circumstance or considerations of context. When interpreted in terms of context, the meaning of a provision may be quite different (often more accurate) from its literal meaning. Thus the true intent of a provision is the meaning carried by language when read in terms of its appropriate context.
- Barring circumstances indicating otherwise, it is presumed that a statute's author has followed established conventions of language, common to the author and audience. Further, it is assumed that the

author has used words in sentences common or appropriate for the subject addressed.

Collectively, these two principles should be used in interpreting NEPA's mandate and regulatory requirements. It is also important to note that the word "shall" is used extensively throughout the NEPA regulations (regulations). Special heed should be paid to use of this term. When used in a legal context, the term "shall" carries the connotation of "must" and is given considerably more deference by the courts than words such as "will" or "should."

10.3.1 What constitutes "reasonable"?

The term "reasonable" is used throughout the regulations. Because "reasonable" is a subjective concept, it is instructive to investigate the meaning of the word in more detail. *Black's Law Dictionary* defines reasonable to mean "... proper ... suitable under the circumstances," "fit and appropriate to the end in view," "... not immoderate or excessive."[3] As a subjective standard, its meaning depends on the specific circumstances at hand and must therefore be interpreted in terms of the prevailing context and conditions.

Reasonably foreseeable: The regulations also shed some light on the term "reasonably foreseeable." As used in the context of §1502.22(b), an impact is considered "reasonably foreseeable" if it is:

1. Not based on pure conjecture.
2. Supported by credible scientific evidence.
3. Within the "rule of reason."

Case law suggests that the threshold for determining if an impact or action is "reasonably foreseeable" is not high. The threshold appears to be lower than that for "likely" or "probable," and is closer to the domain of a "logical possibility." In one case, a court held that the Forest Service must analyze effects of petroleum exploration and development when issuing a lease. This analysis was required even though drilling activity generally occurs in a very small percentage of such leases.[4]

The "rule of reason" standard: An overly strict application of a requirement may result in a decision or course of action that is dangerous, wasteful, inconsistent, ridiculous, or just plain dumb. Consequently, a standard referred to as the rule of reason is applied by courts in reviewing various NEPA issues. In essence, the rule of reason is a judicial mechanism for ensuring that one does not lose sight of reason or common sense. With respect to NEPA, the rule of reason is used in determining whether an EIS is adequate.

10.4 A substantive versus procedural process

Debates persisted for many years after the enactment of NEPA as to whether agency decision makers had a substantive duty to reject environmentally unsuitable proposals. Questions also arose about whether an agency had to be forced to implement mitigation measures. Boggs presents an insight overview of this debate.[5] From a legal perspective, such questions essentially turn on the issue of whether NEPA is procedural or substantive. In examining this issue, we begin by defining specifically what is meant by these two terms

Black's Law Dictionary describes the difference this way:[6]

> ... Laws which fix duties [and] establish rights and responsibilities ... are "substantive laws" ... while those which merely prescribe the manner in which such rights and responsibilities may be exercised and enforced in a court are "procedural laws."
>
> Substantive law: That part of law which creates, defines and regulates rights and duties of parties, as opposed to [procedural law] which prescribes methods of enforcing the rights.

Under a substantive mandate, agencies would have a duty to reject or modify actions that negatively affect the environment. Under a less rigorous procedural mandate, agencies must comply with NEPA's requirements in preparing the analysis but are not obligated to take actions that are environmentally propitious.

From a limited context, NEPA contains some provisions of a substantive nature; Section 101 states that the federal government will use all practicable means of "... attaining the widest range of beneficial uses of the environment without degradation, risk to health or safety, or other undesirable or unintended consequences." NEPA further requires that all practicable means be used to ensure a "... safe, health productive, and esthetically and culturally pleasing" environment. In simple language, if these substantive requirements are enforceable, every agency should be required to make decisions limiting environmental degradation and protecting the environment. This debate is discussed in the next section.

10.4.1 NEPA is essentially a procedural process

While Congress clearly has the power to pass laws governing agency actions, constitutional questions have arisen regarding the power of Congress to pass certain types of laws involving a substantive decision making mandate. In a precedent-setting case, the Circuit Court of Appeals for the District of Columbia ruled in 1971 that in respect to NEPA, courts can reverse a substantive decision made by an agency if it can be shown that the agency's decision was arbitrarily based or clearly did not give sufficient consideration to environmental values.[7]

The Supreme Court, beginning with *Kleppe* in 1976, gave the first indication that it believed NEPA's mandate was largely procedural.[8] This early indication was reinforced in 1978 in the case of *Vermont Yankee*.[9] Later, in 1980, the Supreme Court considered this issue in *Strycker's Bay*, which arose over an urban renewal plan to increase occupancy in a federally subsidized New York housing project.[10] In *Strycker's Bay*, the court concluded that the Department of Housing and Urban Development had appropriately considered environmental concerns in reaching its final decision and that "NEPA requires no more." In 1983, the U.S. Supreme Court heard the case of *Baltimore Gas & Electric Co*. Here the court upheld a rule adopted by the Nuclear Regulatory Commission that assumed that no environmental impacts resulted from storage of certain types of nuclear waste. The court affirmed the commission's right in reaching this decision by declaring that the agency would have violated NEPA only if it had acted arbitrarily in reaching its decision.[11]

Robertson v. Methow Valley: These earlier decisions culminated in 1989 with the historic case of *Robertson v. Methow Valley Citizens Council*, which involved a plan to expand a ski resort in Washington State. In *Methow Valley*, the Supreme Court essentially laid to rest any remaining questions regarding NEPA's mandate to protect the environment. Fogleman (1990) observes:

> ... By stating that an agency complies with NEPA's procedures even though the action is environmentally destructive, the Supreme Court seems to have ruled out any possibility of interpreting NEPA to have any substantive content. ... Of course if the documentation required by NEPA revealed a potential violation of other environmental laws, plaintiffs could base a challenge on those violations.[12]

In the case of *Methow Valley*, the Supreme Court essentially ruled out the possibility of interpreting NEPA to have any substantive mandate in protecting the environment. Although many environmental activists may regret this limitation and the legal community and other environmental professionals have been less than unanimous in endorsing this controversial decision, procedural challenges still offer a potent weapon to those questioning the wisdom of a project. As a result, plaintiffs have refocused efforts on challenging environmental considerations from a procedural perspective.

10.5 Requirements for bringing a successful environmental lawsuit

With respect to NEPA litigation, a plaintiff is the party challenging an agency's action and the defendant is the agency against which a challenge is mounted. Environmental activist organizations and citizen groups have been the most common plaintiffs. The NEPA statute does not explicitly provide for citizen suits, so standing is most commonly established through the Administrative Procedure Act (APA).[13]

The term "jurisdiction" is the authority of a court to hear a case and render a judgment. In NEPA litigation, this usually involves demonstrating that the issue before the court is based on a federal question; the question is whether the agency's action complies with NEPA's requirements. Several requirements must be met before a court can hear a case. These criteria are briefly summarized in the following sections.

10.5.1 Ripeness

In general, an agency must first be given an opportunity to complete its NEPA process. Judicial review of an agency's compliance with NEPA is not to begin until the agency:

> ... has filed the final environmental impact statement or has made a final finding of no significant impact ... or takes action that will result in irreparable injury (§1500.3).

If an action, subject to NEPA, is taken that could result in "irreparable injury," prior to issuing a FONSI or ROD, a plaintiff may request the court to grant an injunction, halting the agency's action while the court considers the case. Thus in seeking judicial resolution, the cause of action must be ripe; that is, the agency has determined that NEPA does not apply or has not followed NEPA's procedural requirements. Otherwise, a challenge must wait until the NEPA process has been completed.

Prior to seeking judicial resolution, a plaintiff must have exhausted any available administrative remedies. For example, the right to bring a suit may be denied if the plaintiff (1) fails to comment on the agency's proposed action; (2) fails to use the agency's administrative review procedures; or (3) remains silent on NEPA issues until such time that a suit is filed.

Neither the regulations nor the NEPA statute contain a statute of limitations. However, suits may be barred by a legal restriction referred to as latches. Under the latches doctrine, courts may bar a suit if the agency can show that the delay in bringing the suit was inexcusable.

Trivial violations: The CEQ has stated that it does not believe that unnecessary suits should be generated as a result of NEPA. During public review of the draft regulations, concerns were raised that agency actions might be invalidated in litigation proceedings as a result of trivial departures from regulatory procedures. Consistent with this position, CEQ modified the draft regulations, adding the following provision: "any independent cause of action" (§1500.3).[14] Thus a trivial violation should not by itself give rise to a lawsuit (i.e., independent cause of action). If, on the other hand, a nontrivial error is made, a suit may encompass both trivial and nontrivial violations. Of course, triviality is a subjective consideration that, much like significance, requires considerable professional interpretation.

10.5.1.1 Criteria for "standing"

Specific criteria must be met before a case can be presented before a federal court. If these criteria are met, a plaintiff has standing (i.e., grounds) to bring a case before the court. Standing indicates that a party bringing a suit has sufficient stake in the agency's action, such that a court's decision will address the grievance. The courts tend to place only a limited set of few restrictions on individuals and organizations who may file an NEPA suit. In general, plaintiffs must show only that they have more involvement in the issue than a general concern in defending the environment.[15] An Injury in Fact Test has been applied in determining whether a plaintiff has standing to bring a suit, based on NEPA.

The Injury in Fact Test: Consistent with the APA, an Injury in Fact test has been applied in determining whether a party has standing.[16] In respect to NEPA, the APA establishes four primary conditions that must be met before a plaintiff has standing to bring a suit. The plaintiff must demonstrate that:[17]

1. The plaintiff could be harmed by the agency's action.
2. Either a potential or actual injury would be caused by the agency's action that is not in compliance with NEPA.
3. The injury could be alleviated by a favorable decision.
4. The potential injury involves environmental issues and not simply economic ones.

The purpose of NEPA is to protect the environment. NEPA does not protect economic interests (although socioeconomic interests such as environmental justice and impacts on community services do fall under the purview of NEPA). Thus as indicated by the fourth item in the preceding list, a suit motivated exclusively by economic concerns does not have standing. However, a party may have standing if the suit is motivated principally, but not solely, by economic interests.

As depicted by the first and second items listed above, claims involving imaginary circumstances in which the plaintiff could be affected, a general interest in preserving the environment or a generalized fear of losing an environmental resource is insufficient to show standing. To be adversely affected, a plaintiff must be within the "zone of interest" for which protection is sought.[18] In 1989, the U.S. Supreme Court appears to have made conditions necessary for demonstrating standing more difficult. Here, the court ruled that a plaintiff could not obtain standing simply by arguing that it uses certain lands within "the vicinity" of the potentially affected areas. Although potential injuries must be identifiable, they need not be extensive. As a rule, an individual or organization must show only that it derives some benefit from use of the adversely affected environment. Where the plaintiff is an organization, it need only demonstrate that some of its members benefit from use of the resource. The reader is referred to Mandelker for a more thorough discussion of standing and the Injury in Fact Test.[19]

10.5.1.2 Mootness

A plaintiff's interest in a challenged action must continue over the course of the litigation process. Actions taken by the agency to rectify the legal problem may cause the plaintiff's challenge to become moot (i.e., lack standing). For example, a challenged action may become moot if the agency takes steps to bring itself into compliance with NEPA. A second example includes circumstances in which an action becomes infeasible and the agency cancels the proposed action. Mootness does not apply in cases where there is a threat of a future NEPA challenge to a project that is likely to continue.

10.6 Litigation process

Courts have generally not ordered agencies to comply with NEPA where an action has been completed and the agency has acted in good faith. An instance in which an agency has not acted in good faith is another matter. An agency can be successfully challenged for failing to fulfill commitments made in an EIS and ROD or EA and FONSI (usually a mitigated FONSI) or for undertaking actions not evaluated in the chosen alterative. The following sections describe legal principles, concepts, and strategies fundamental to the understanding of NEPA litigation.

10.6.1 Enforcement of NEPA

The scope of NEPA's regulations is "... applicable to and binding on all federal agencies for implementing the procedural provisions of the Act" (§1500.3).

The regulations cite only one exception to this provision, which involves circumstances where "compliance would be inconsistent with other statutory requirements" (§1500.3). The courts have taken a very narrow interpretation of this exception. Federal agencies bear primary responsibility for enforcing NEPA, with the courts intervening only when a plaintiff charges that an agency has failed in its duty. Specifically:

> The President, the Federal agencies and courts share responsibility for enforcing the Act so as to achieve the substantive requirements of section 101 (§1500.1[c]). In reality, the president rarely, if ever, becomes involved in such enforcement issues, other than through the Council on Environmental Quality (CEQ), which is an executive office representing the president.

NEPA contains no explicit provision for enforcing NEPA. Nor has CEQ or the EPA been granted enforcement authority beyond that of the referral process. Although NEPA does not expressly provide for judicial review, the courts have held that such review is implied.[20] Consequently, the courts, by way of lawsuits, have become the de facto mechanism for enforcing NEPA.

10.6.2 Legal system

Challenges based solely on NEPA issues are not tried before juries. A legal challenge normally begins in a U.S. district court. District court decisions may be appealed to the U.S. Court of Appeals for the particular district. If the court agrees, a case may be elevated to the level of the U.S. Supreme Court.

The district court acts as a fact finder and the role of the appellate court tends to be one of reviewing the district court's record. Thus an appellate court's scope of review is more limited than that of the district court. Because the appellate court normally limits its review to the factual material presented in the lower court's record, it grants a substantial degree of deference to the district court's finding of fact. A lower court's finding is not overturned unless it clearly lacks a reasonable basis or is legally flawed.

Venue: The term venue refers to the location and court in which a suit will be heard. Cases involving NEPA are filed in federal court. A case can be brought before any district court whose jurisdiction covers the location of the "cause for action" or where the agency conducts business. Typically, these cases are heard in the district court having jurisdiction over the location in which the federal agency's action would take place. Because many agencies are headquartered in Washington, D.C., cases are often brought before the U.S. District Court of the District of Columbia.

Frequently, a plaintiff often has a choice of several federal courts in which to file a suit. A smart plaintiff will screen the possibilities with care, inasmuch as each federal judicial circuit has its own body of case law, which may work either in favor of the plaintiff or to the plaintiff's detriment. For example, the U.S. Forest Service can be sued in (1) the District of Columbia, where the chief's office is located; (2) the state in which the regional forester's office is located; (3) the state in which the Forest Services supervisor's office is located; or (4) the state in which the project is located.

10.6.3 Bringing a suit against an agency

Where a party is unsatisfied with a final decision, it may challenge an agency's NEPA process. The court's ability to hear an environmental case begins when plaintiffs demonstrate that they are the proper individuals to challenge the federal action and that the issues they are raising are issues that the court is authorized to consider. This is known as "standing," which is discussed in more detail later in this chapter. That certain issues are suitable for judicial resolution is known as "reviewability." Virtually all issues that may arise in an environmental law case are subject to judicial review.[21]

Filing a suit: Some agencies have an administrative appeal procedure to which potential plaintiffs can take their concerns. An agency's appeal procedure must be exhausted before the judicial process can be pursued.

The procedural requirements that must be followed by both parties are specified in the Federal Rules of Civil Procedure. The plaintiff begins by filing a complaint. The complaint is a short statement explaining why the agency should take the action the plaintiff is requesting. In response, the defendant (the agency) may file a reply in which it defends its position. The agency may also file a motion to dismiss if the plaintiff fails to demonstrate an adequate remedy to the agency's action. Either party may make a motion for summary judgment in which the court renders its final decision. A summary judgment is made only when the issue is confined to a question of law. A settlement agreement is often reached to avoid a costly and protracted suit, and is normally sought when both parties basically agree on the action to be taken.

Discovery: Where a suit involves an issue of adequacy, the plaintiff typically makes a request to examine the NEPA documents and other supporting sources of information. This step is referred to as discovery. The agency provides the plaintiff with a copy of its administrative record. Plaintiffs may also seek discovery beyond the administrative record; however, the courts normally restrict discovery to the administrative record unless there is reason to believe that it is incomplete.

Class action suits: A case tends to be strengthened by demonstrating that an action could harm a large number of people. A class action suit provides a mechanism for the "man on the street" to exploit such strength. Although they provide a potent platform for challenging the massive resources of a federal agency, class action suits are not without their own risks.

Prudence should be exercised in defining an appropriate class. For example, the courts have not looked favorably on class action suits in which a potential action may affect the group in conflicting ways. If a class action is too large or diverse, the court may find that the group cannot be properly identified or represented, or that its members have conflicting interests. Conversely, if the class is too narrowly defined, the court may dismiss the suit for failing to incorporate potentially affected members. In one case, a class action suit was brought on behalf of farmers who would be affected by a highway construction proposal. The court found that there was a conflict between those farmers who would be harmed by construction of the highway and those who would be harmed if an alternate route were chosen.[22]

10.6.4 Legal strategies

As stated earlier, the U.S. Supreme Court established a key precedent by ruling that once an agency has complied with NEPA's procedural requirements, it may proceed with an action even though that action maybe environmentally destructive. For this reason, plaintiffs have refocused their efforts on challenging procedural violations. Although adversaries may find this limitation regrettable, procedural challenges still offer a formidable weapon for opposing an agency's action. For example, if a potentially significant environmental issue has not been addressed or has been inadequately investigated, procedural grounds exist for challenging the agency. Of particular note to readers of this book, if an action with potentially significant environmental action were addressed through an EA rather than EIS, procedural grounds would exist for a challenge.

Plaintiff's strategy: Actions such as the mere threat to pursue legal action often provides impetus sufficient to pressure an agency into reconsidering its proposed action. Circumstances permitting, such a threat can provide adversaries with influence sufficient to exercise some degree of influence in shaping federal policy. If the plaintiff successfully argues its case before the court, a project may be halted and the agency may be compelled to initiate or repeat the NEPA process.

To increase the prospect of success, plaintiffs should be able to demonstrate that the possible environmental impacts affect the public, or a large segment of it, in addition to themselves. Plaintiffs should also focus on demonstrating that the action would result in environmental degradation beyond simply economic losses. Plaintiffs are advised to consult experts intimately familiar with all aspects of NEPA compliance in an effort to identify legitimate procedural grounds.

Normally, plaintiffs ask for a preliminary injunction. The decision to grant or deny an injunction lies with the court and may be appealed. In reaching a decision to grant an injunction, the court considers the specific circumstances of each case. Factors considered by the court include the merits of the plaintiff's claim and the environmental harm that would result from not granting an injunction, balanced against the injury to the agency should the project be halted.

Agency's strategy: Depending on the circumstances, a plaintiff's strategy may not be to win, but simply to delay a project to the point where the defendant (agency) capitulates. Prudent planning necessitates that the agency consider such a possibility and take measures to reduce its vulnerability. In countering such tactics, all reasonable alternatives and mitigation measures should be thoroughly investigated and seriously considered. Experience has shown that a cooperative relationship can be highly effective in reducing the controversial nature of the action and diffusing the risk of litigation. This underscores the importance of careful public involvement planning and outreach to interested parties throughout scoping and other elements of the NEPA process, as well as why scoping efforts that may be optional for an EA and FONSI might still be advantageous to an agency.

From the agency's perspective, every effort should be taken to ensure that NEPA requirements have been rigorously complied with. Even for an EA, an auditable trail should exist so that if challenged, the agency can clearly and swiftly demonstrate compliance to the court. Some of the pertinent aspects underlying the process used in filing NEPA lawsuits are described in the following paragraphs.

10.6.5 Court's role

Once plaintiffs establish standing and reviewability, the issue becomes the extent to which the court is willing to second-guess the agency as opposed to deferring to that agency based on its expertise. Judicial review depends on the kind of issues involved in the case; that is, whether they are issues of law or fact. If the issues are questions of law, the court will not necessarily feel bound by the administrative agency's interpretation of the law. If the issues are questions of fact, the standard of review depends on the type of proceeding.

Where an action involves an adjudicative action, if the agency shows substantial evidence in the record to support its ruling, the action is upheld. For nonadjudicative administrative decisions, the standard of review is an "arbitrary and capricious" test, described in a later section. To apply this test, the court examines the administrative agency's record. The court decides whether the agency considered all the relevant factors and whether the agency made a clear error in judgment.[23] Some courts apply the arbitrary and capricious test rigorously, scrutinizing the reasons given for the administrative decision to determine whether adequate support exists within the record. This has become known as the "hard look" approach, which is discussed later in this chapter.

10.6.6 Legal remedy

Where a court has found that an agency violated an environmental law, the plaintiff may seek a remedy. Sometimes this is only an order for the agency to give further consideration to the issue before taking agency action. Under NEPA, remedies usually include an injunction to take or not to take some sort of action by a certain time or after certain other actions have been undertaken. NEPA remedies are explained later in this chapter.

10.6.7 Reviewing an agency's administrative record

With few exceptions, courts generally confine their review to the agency's administrative record.[24] The importance of establishing a comprehensive administrative record, demonstrating that the agency properly weighed all pertinent factors in reaching a final decision, cannot be overemphasized. In one case, the Natural Resources Conservation Service was challenged for preparing an inadequate EIS, which led to approval of a permit by the U.S. Army Corps of Engineers (Corps) for construction of a dam that could affect a number of species, two of which were endangered. The Corps argued that it had relied on its internal experts in determining that there was no additional evidence indicating that the EIS must be supplemented. The court rejected the Corps' argument, as the administrative record failed to demonstrate that the staff members involved in the review and analysis were qualified to address this issue.[25]

What constitutes an adequate administrative record: It is difficult to precisely define exactly what constitutes an agency's administrative record, because NEPA does not establish a formal fact-finding process. For the purposes of judicial review, an agency's administrative record normally includes the entire record that existed at the time the decision was made and not simply the portion of the record read by the decision maker.[26] At a minimum, the administrative record should include "... all relevant studies or data used or published by the agency."[27]

Generally, an administrative record should consist of all documents relied upon in preparing the NEPA document, as well as those considered by the decision maker. The record should include public notices, references, technical studies, appendices, concurrences, public comments and responses, and draft copies of the NEPA analyses. Additional documents should include, but are not limited to, hard copies, electronic files, calculation sheets, field notebooks, photographs, telephone conversation records, comment reviews, exchanges in opinions, meeting minutes, and internal and external memorandums. The purpose of the administrative record should be to demonstrate that the agency really took a "hard look" at the proposed action and reasonable alternatives (and did not just say they did).

Of particular note, e-mails can become part of the administrative record. This is particularly true for e-mails containing technical information or technical analysis or interpretation. This underscores the need for authors of e-mails to assume that the contents of their messages will become available to a public audience. Environmental professionals must ensure that professionalism underlies each and every e-mail they send, even purely social ones that lack technical content. Drives containing e-mail records of agencies, contractors, plaintiffs, and other interests can be subpoenaed by courts. With today's computer equipment no computer data, including e-mails, can ever truly be deleted with complete confidence. Sophisticated information technology experts can easily extract a lot of seemingly "deleted" communications.

On the other hand, while e-mails can be used as a way of storing technical information in an administrative record, they are not the best. It is always good professional practice to ensure that key technical information from e-mails is transferred to a more traditional medium such as a memorandum or report.

Delving beyond the agency's administrative record: Normally, it is imprudent for judges to consider testimony and documentary evidence unless this material was first presented and considered by the agency in question. Nevertheless, a court cannot meet its responsibility in conducting a "substantial inquiry" if it must simply accept an agency's word that it considered all relevant matters.

In some narrow circumstances, the court must look beyond the administrative record to verify that the agency considered all relevant factors. As stated earlier, only in limited circumstances will a court expand its review or permit discovery beyond the agency's record.[28] Allegations that an EIS or EA has not investigated reasonable alternatives, overlooked significant impacts, or swept "stubborn problems or serious criticism ... under the rug" may raise questions sufficient to justify introduction of new evidence outside the administrative record, including expert testimony. Such evidence may be introduced where challenges involve either the adequacy of the analysis or cases involving an agency's finding that an EIS is not required.[29]

In scrutinizing the reasoning behind an agency's decision, a court may also find it necessary to inquire outside the administrative record by introducing affidavits and testimony. Such review is also appropriate if there is reason to believe that an agency used other material outside the record, acted improperly, or acted in "bad faith" in respect to its knowledge.[30]

Inquiries that proceed beyond the administrative record should be limited to collecting background information. As such, inquiries should be limited to the purpose of "... ascertaining whether the agency considered all the relevant factors or fully explicated its course of conduct. ... Consideration of the evidence to determine the correctness or wisdom of the agency's decision is not permitted, even if the court has also examined the administrative record."[31]

In general, a court cannot undertake its own examination of a decision after concluding that an agency's inquiry was inadequate or is not supported by the administrative record. Once the court has concluded that the agency acted inappropriately, the court's duty is to return the matter back to the agency for additional examination.[32] Thus the court's role is to determine whether the authors of an EIS or EA have conducted "an adequate and objective effort, judged in light of the 'rule of reason,' to compile and present all significant environmental factors and alternatives for the decision maker's consideration."[33]

Expert witnesses: In certain restricted instances, scientific evidence beyond the administrative record is allowed to be introduced. Expert testimony, which can be oral or written, is often used for introducing such evidence. Both authors of this book, geologist Charles Eccleston and ecologist Peyton Doub, have served as expert witnesses. Such testimony can be admitted where the expert is adept in a particular field by way of "... knowledge, skill, experience, (or) training."[34] Publication of scientific "... treatises, periodicals, or pamphlets" can provide a basis sufficient to allow an expert witness to be cross-examined.[35] Courts may also admit government reports and findings from an "... investigation made pursuant to authority granted by law."[36]

10.7 Legal standards applied in reviewing NEPA

The NEPA statute does not prescribe a judicial standard to be used in legal reviews. Partly for this reason, the applicable legal standards for reviewing various issues have been unclear and the subject of conflicting judicial decisions. The situation changed, however, beginning with the historic case of *Overton Park*.[37] Legal standards applied by the courts in reviewing NEPA cases are described in the following sections. These standards vary, depending on the specific issue at hand.

10.7.1 Administrative Procedures Act

Since administrative law issues are often involved in environmental regulatory disputes (including NEPA), it is instructive to consider a brief overview of the Administrative Procedures Act (APA). Agency conduct is governed by the APA. Under the APA, agencies can act only pursuant to the authority granted to them in statutes by Congress. Agencies are mandated to follow their own rules and regulations.

A court's review of an agency's decision making process is also governed by the APA. When a federal agency's decision is in question, the issue is not whether the decision was good or bad but, instead, whether the agency correctly followed procedures and was within its statutory authority in reaching the decision. Agencies may not act or make decisions in an arbitrary or capricious manner and must maintain a record of evidence indicating that procedures were correctly followed. Courts grant a great degree of deference to agencies in interpreting their own regulations and the statutes that they administer.[38]

10.7.2 Judicial review standards

The APA provides judicial review standards for a wide variety of types of agency decision making.[39] In respect to judicial review standards, the APA dictates that courts (this does not apply to adjudicative decisions) are to set aside agency actions and conclusions that are found to be:

- Arbitrary and capricious, an abuse of discretion, or that are not in accordance with the law.
- Contrary to constitutional power and privilege.
- In excess of statutory authority or jurisdiction.
- Without observance of procedures required under law.

10.7.2.1 Disagreement among experts

Federal agencies routinely deal with complex technical issues; but judges typically lack such expertise. To complicate matters even further, it is not uncommon for scientific experts to disagree on issues of a technical nature. In cases where an agency's expertise and judgment are questioned, courts have typically sided with the agency. One bench went so far as to state that the court:

> ... (Is not in the) same position as the agency in its review of the scientific data submitted, and cannot replace the agency's judgment with its own.[40]

The Supreme Court, for example, considered an agency's decision not to prepare a supplemental environmental impact statement (S-EIS) in 1989.[41] Differences existed within the scientific community in respect to the significance of new information that had been obtained. Because the case involved professional disagreement among experts, the court deferred in favor of the agency's experts. The court concluded that, based on its internal expertise, the agency's experts and not the plaintiff's experts were in the best position for determining which view to believe.

10.7.3 Court's role

While the Supreme Court deferred in favor of the aforementioned agency's technical expertise, it cautioned that courts should not automatically do so. This ruling appears to indicate that the court's role is to first review the administrative record and determine whether the agency made a sound decision based on available information. The court's role, however, is not to second-guess the professional judgment of an agency in such matters. Instead, its role is to determine whether the agency can demonstrate that it carefully reviewed the facts before reaching a decision.

Disputes involving an agency's discretion to interpret laws and regulations are another matter. Such matters normally turn on one's ability to make legal interpretations, in contrast to the ability to assess technical and scientific facts. Because judges have considerable expertise in interpreting legal requirements, the courts have taken a less deferential view toward agencies. With one exception, the courts have not been reluctant to substitute their own judgment for that of the agency's when matters involved legal interpretation. In respect to the exception, courts generally give a considerable degree of deference to an agency in the "reasonable" interpretation of its own regulations. The courts have also given agencies considerable discretion in interpreting the CEQ NEPA regulations.[42]

10.7.3.1 Determining whether an EIS is required

The principally cited reason for NEPA lawsuits has simply been failure to prepare an EIS. This might consist of simply proceeding with an action without completing any NEPA review, or proceeding with an action involving significant environmental impacts on the basis of a categorical exclusion or EA and FONSI. Obviously, the standard used for reviewing such cases is of utmost concern. Until recently, the courts were divided regarding the appropriate standard. Some courts applied an arbitrary and capricious standard, which tends to favor the agency. Other courts have favored a reasonableness standard, which is less deferential to the agency. Beginning with the historic case of *Overton Park*, this issue has been the subject of wide and often conflicting litigation.

Important Supreme Court decisions: The case of *Overton Park* arose over a federal law prohibiting construction of highways through public parks unless the secretary of transportation determined that there was no feasible alternative.[43] The secretary was challenged, after determining that no feasible alternatives existed. The case made its way to the Supreme

Court, which was faced with having to determine the appropriate judicial standard to be used in reviewing the secretary's decision.

As described earlier, the APA places constraints on the degree to which courts may overturn agency decisions. The Supreme Court concluded that the arbitrary and capricious standard of the APA was "generally applicable" in reviewing an agency's decision not to prepare an EIS. The court went on to state that courts are to engage in a "substantial inquiry" in applying such a standard. In respect to their judicial responsibility, the Supreme Court stated that courts must determine whether a decision has been based on review of:

> ... (The) relevant factors and whether there has been a clear error in judgment. ... Although this inquiry into the facts is to be searching and careful, the ultimate standard of review is a narrow one. The court is not empowered to substitute its judgment for that of the agency.[44]

This view was again reinforced by the U.S. Supreme Court in 1989, which ruled that a court is not to set aside an agency's decision unless there has been a "clear error in judgment."[45] Absent a finding that the agency's decision was arbitrary and capricious or an obvious abuse, the court is not to substitute its judgment or second-guess an agency's decision once it is satisfied that serious consideration has been given to the EIS analysis.[46]

The "arbitrary and capricious" standard: With respect to NEPA, "arbitrary and capricious" means that an agency's decision making process was based on factors Congress did not want considered, or that the agency completely failed to consider important and relevant aspects to the issue in question. Specifically, the court's role is to determine:[47]

- Whether an agency's decision was based on a consideration of the relevant factors.
- Whether there has been a clear error of judgment.

Under the arbitrary and capricious standard, the agency's administrative record must contain sufficient evidence to support its decision not to prepare an EIS. A court is apt to find that an agency acted in an arbitrary and capricious manner if it offers an explanation that is:[48]

- Contrary to the evidence.
- So implausible that it cannot be attributed to differing views or expertise.

This arbitrary and capricious standard may be applicable to other NEPA legal issues as well. The "substantial inquiry" standard is closely akin to the "hard look" doctrine, which has been widely applied to the review of other environmental statutes; these terms are discussed shortly.

10.7.3.2 Determining if an EIS needs to be supplemented

In the case of *Marsh*, the Supreme Court held that the arbitrary and capricious standard is to be used in reaching a decision not to prepare an S-EIS. In rejecting the reasonableness standard, the court noted that the actual difference between the two standards is modest.[49] Although the court rejected the reasonableness standard, it appears to have endorsed the hard look doctrine, described shortly.

When must an EIS be supplemented? An agency is not required to supplement an EIS every time a change occurs or new information becomes known. Indeed, courts have concluded that the rule of reason is applicable in making such determinations. In *March*, the Supreme Court ruled that a decision to prepare an S-EIS should be based on a methodology similar to that used for making a decision to prepare an EIS in the first place.[50] At least two criteria have been specifically identified by the courts, which trigger the requirement for supplementing an EIS:

- The action remaining to be completed (e.g., not already covered in a NEPA document) must meet the definition of a "major federal action." The second criterion need be considered only if the first criterion is met.
- There is new information, indicating that the remaining action could result in (1) significant environmental impacts not originally documented or (2) impacts substantially different from those originally documented.

In *March*, there were conflicting views within the scientific community regarding the significance of new information. Because the case involved professional disagreement among experts, the court deferred in favor of the agency.[51] Although the Supreme Court deferred in favor of the agency's technical expertise, it cautioned that courts should not automatically do so. The court's role is to first review the record and determine whether the agency made a sound decision in respect to the significance of the new information.

An agency must demonstrate that it reviewed the facts carefully before concluding not to supplement an EIS. The reader should note that the regulations also require that an EIS be supplemented where an agency makes substantial changes in the proposed action or where significant new circumstances arise (§1502.9fc]). It is recommended that decision makers consult with NEPA specialists (preferably also technically conversant in the specific technical issue(s) involved) and legal counsel in reviewing the need to supplement an EIS.

What constitutes an adequate analysis? After the issue of failing to file an EIS (or EA), the second most citied complaint in NEPA suits involves claims that an EIS (or EA) is inadequate. Mandelker defines adequacy to mean "... whether an agency had complied with the environmental review process mandated in NEPA by preparing an impact statement that adequately considers the environmental effects of the proposed action."[52]

In challenging the adequacy of an EIS or EA, a plaintiff is normally required to provide only a reasonable argument, not undisputable evidence (i.e., *prima facie* evidence), that the document is inadequate. Once this has been done, the agency bears the burden of proof in demonstrating that the document is adequate.[53]

Consistent with the "rule of reason," Eccleston has developed a tool for determining when an analysis contains information sufficient to satisfy the requirements of NEPA.[54] The methodology used by the courts in determining adequacy is described in the following sections.

The "hard look" doctrine: A doctrine known as "hard look" has often been cited by courts in their review of the adequacy of an EIS. General applicability of the hard look doctrine is still unclear, although the Supreme Court appears to have come close to endorsing its use. This doctrine is not an actual review standard in itself. Rather, it is a legal mechanism for implementing certain standards that the courts have determined to be appropriate. Essentially, the hard look doctrine dictates that a more rigorous review of agency actions and decision making is to be conducted when environmental issues are at stake. As espoused by Mandelker, this doctrine dictates, "courts must undertake a hard look to ensure that the agency took a hard look."[55] In the words of one court:

> ... (A)ssumptions must be spelled out, inconsistencies explained, methodologies disclosed, contradictory evidence rebutted, record references solidly grounded, guesswork eliminated and conclusions supported in a manner capable of judicial understanding.[56]

Legal counsel should be consulted in respect to questions regarding its applicability.

10.7.3.3 *Agency's final decision*

On completing the EIS, a decision maker is required to review and understand the statement before reaching a final decision, which is then recorded in the ROD. The courts have indicated that an agency has not fulfilled its NEPA mandate by preparing an EIS merely to satisfy a procedural requirement and then completely disregarding the analysis in its subsequent decision making process.[57] Such practice is viewed as arbitrary and capricious because the final decision is detached from the NEPA analysis. An agency must be able to demonstrate that results of the NEPA process were seriously considered and balanced against other competing factors before a final decision was reached. To this end, the court's role is to conduct a "substantial inquiry" to ensure that the agency has taken a hard look at the issues analyzed within an EIS before reaching its final decision.[58]

10.7.3.4   What the courts are looking for

With respect to the issue of adequacy, courts focus their review on determining whether an EIS or EA complies with NEPA's statutory provisions. The analysis must be complete and capable of standing on its own merits.[59] The EIS or EA must, in and by itself, either meet the requirements of NEPA or fail the test of adequacy. In essence, an EIS or EA is considered inadequate if it fails to provide sufficient foundation for making decisions based on environmental factors.

With the except of major deviations to Section 102(2)(C) of NEPA, the basis for determining adequacy defies summary. Although agencies must normally take a hard look at the consequences of potential actions, the courts are not looking for perfection, the impossible, or objection-free documents. Consistent with the rule of reason, courts have held that agencies are not expected to use a "crystal ball" approach in their analysis. An EIS need not be "exhaustive," considering a problem "from every angle." Nor do courts normally sustain criticism that is considered "overly technical," "hypercritical," or indulges in "chronic faultfinding."[60] Similarly, an EA need not exhaustively analyze every conceivable possibility of significant impacts, including those with only a remote likelihood of occurring. However, the bar for not considering an issue in an EA is generally greater than excluding it from an EIS; the EA was never intended to enable agencies to circumvent the rigor of an EIS for actions that could indeed result in significant environmental impacts. Note that EAs are intended for relatively simple and noncontroversial actions without a potential for significant environmental impacts; EA preparers anguished with a torrent of issues requiring extensive thought to defend nonsignificance should consider whether an EIS would be a better approach.

Characteristics of adequacy: Reviewing courts are likely to be less demanding in scrutinizing EISs or EA/FONSI documents where the effects of an action are confined "in scale" or involve "no unusual environmental effects." This also holds true where "effects are more social than environmental," the analysis is "remote and speculative," the outcome of the decision is "practically preordained," or where there is little room for variation in the alternatives. In contrast, the courts have been particularly demanding where actions are incompatible with land use planning restrictions, exact a heavy toll on the environment, trigger other connected actions, or have been prepared in secrecy.[61]

Table 10.1 Characteristics of inadequate analyses and inadequate analyses

Adequate characteristics	Inadequate characteristics
Presents information sufficient to enable the decision maker to consider the environmental effects and make a reasonable decision[64]	Contains discussions or conclusions that are:
	internally contradictory
	basically flawed
Discusses:	excessively cryptic
connected actions[65]	argumentative
significant aspects of the probable consequences thoroughly and objectively[66]	genuinely preposterous
	dependent on outdated information or biased procedures
reasonable direct and indirect impacts[67]	misleading or incomplete
	wholly unquantified
cumulative impacts[68]	vague
synergistic effects[69]	unsupported by fact
Presents more than a mere listing of alternatives[70]	scientifically indefensible
	sweeping
	incomprehensible
	Ignores important topics or exudes arrogance
	Demonstrates a reluctant or begrudging compliance

Experience indicates that an inadequate EIS or EA is often characterized by conclusions that are "scientifically indefensible," "obviously misleading," "basically flawed," or "argumentative" or "ignore important topics" or demonstrate reluctant compliance. Specific areas that have been vulnerable to a finding of inadequacy include insufficient description of the proposed action or alternatives. Insufficient analysis of environmental impacts, especially in terms of indirect and cumulative impacts, has also frequently resulted in a finding of inadequacy. Other flaws include failure to discuss adequately the relationship between short-term benefits and long-term productivity, and irreversible and irretrievable commitments of resources.[62] Table 10.1 summarizes characteristics that have been found to be adequate versus those that have been judged inadequate. Rodgers provides an in-depth review of the inadequate characteristics summarized in Table 10.1.[63] The reader may want to consult with NEPA specialists or legal counsel regarding more recent developments in case law.

10.8 Judicial remedies in NEPA suits

Plaintiffs can sue a federal agency to enforce NEPA's mandate to adequately "consider" environmental factors in the decision making process. In respect to NEPA, relief is intended for remedial, rather than punitive, purposes. Thus relief is intended to advance NEPA's goals, not to "punish" an agency.

Where potentially significant impacts cannot be discounted, the court may order an agency either to prepare an EIS instead of using an EA or categorical exclusion. When an EIS contains key technical mistakes or deficiencies, the court may order an agency to either prepare a new EIS or correct deficiencies in the existing statement. If an agency "plans" to prepare a new or revised EIS, but a plaintiff charges that there has been an unreasonable delay in doing so, the court may order the agency to begin its preparation. In some cases, the court may even set compliance schedules for completing the EIS. Courts may also retain jurisdiction to review the completed statement.

10.8.1 Recovering legal costs

The act provides no provision for awarding transaction costs (e.g., attorney and legal fees) to plaintiffs. Nevertheless, in many circumstances, a prevailing plaintiff may be eligible to recover such costs. The Equal Access to Justice Act provides a mechanism by which plaintiffs can recover transaction fees against the United States.[71] Plaintiffs are entitled to reimbursement if they prevail in a suit and meet certain eligibility requirements. Judgments are restricted to plaintiffs possessing limited resources. Awards may include reimbursement for litigation expenses such as expert witnesses, scientific tests, and studies.

10.8.2 Types of remedies

Four types of judicial remedies are commonly sought by plaintiffs:

- Preliminary injunction. An order by the court to enjoin (e.g., perform or halt) certain activities until the agency has complied with NEPA.
- Permanent injunction. A court order to enjoin certain activities until the court has reached a final decision regarding the suit.
- Declaratory judgment. The court determines the legal obligations of the agency without requiring proof of irreparable harm necessary for a preliminary injunction.
- Writ of mandamus. A writ of discretion is issued by the court, ordering an agency to perform some duty required as part of NEPA compliance.

Injunctions: In filing a suit, plaintiffs typically seek a preliminary injunction. Three principal types of injunctive relief are available: temporary restraining orders, injunctions, and stays. In respect to NEPA, an injunction is a legal device used to prevent an agency's action from continuing to the point where compliance would be meaningless because it would be too late to halt or stop serious environmental degradation. When a court grants a preliminary injunction, it normally enjoins or prohibits further work on a project, until such time that the agency has adequately complied with NEPA. Some activities, however, may be allowed to continue if they do not result in environmental degradation. Where more serious infractions are alleged, a court may go so far as to order cancellation of a project.

A court may decide not to issue an injunction even though an agency has clearly violated NEPA. Criteria used in granting a preliminary injunction vary with the specific court circuit in which the suit is brought; however, four criteria are generally considered. To be granted an injunction, the plaintiff must show:

- An ample likelihood of success based on the virtues of the case.
- A considerable threat of irrevocable injury if the injunction were denied.
- That the potential injury to the plaintiff outweighs the potential injury that an injunction would cause to the defendant.
- That the injunction would not impair public interests.

Injunctions and their misuse: Challenges are usually based on allegations of agency noncompliance with NEPA. In one case, however, the U.S. Forest Service requested an injunction to prohibit people from using an area of a national forest until the Forest Service had complied with NEPA. The court questioned the Forest Service's motives in requesting this injunction. In ruling against the Forest Service, the court explained that NEPA is not to be used to "... suppress First Amendment activity, or out of hostility to a particular group."[72]

10.9 Recent case law involving NEPA

Swartz preformed a review of substantive NEPA cases issued by federal courts in 2005.[73] This study found that federal courts issued 20 substantive decisions involving NEPA by 10 different federal departments or agencies. The federal agencies prevailed in 13 of the 20 cases (65 percent). As has been the case in previous years, the U.S. Forest Service was the individual agency involved in the most number of cases (6); the Forest Service prevailed in only 2 of the 6 cases. This study identified the following themes:

- Courts wanted to see evidence of meaningful public involvement for EAs:
  - *Sierra Nevada Forest Protection Campaign v. Weingardt* (E.D. Cal)
  - *Alliance to Protect Nantucket Sound, Inc. v. U.S. Department of the Army* (1st Cir.)

- *El Dorado County v. Norton* (E.D. Cal.)
- *TOMAC v. Norton* (D.D.C.)
- Courts invalidated NEPA analyses that relied on flawed data:
  - *Natural Resources Defense Council v. U.S. Forest Service* (9th Cir.)
  - *Native Ecosystems Council v. U.S. Forest Service* (9th Cir. August)
  - *Ecology Center v. Austin* (9th Cir.)
- A court reiterated that a plaintiff that has only an economic interest in a project does not have standing to bring an NEPA case:
  - *Ashley Creek Phosphate Co. v. Norton* (9th Cir.)
- Courts upheld NEPA documents that considered arguably unrelated actions separately as long as the cumulative impacts where addressed:
  - *Methow Forest Watch v. U.S. Forest Service* (D. Or)
  - *El Dorado County v. Norton* (E.D. Cal.)
- Courts reiterated that a cumulative impact analysis need not consider future actions that are too speculative:
  - *City of Shoreacres v. Waterworth* (5th Cir.)
  - *City of Riverview v. Surface Transportation Board* (6th Cir.)
  - *City of Oxford v. Federal Aviation Administration* (11th Cir.)
- A court invalidated an NEPA document that considered related actions separately because, among other things, it did not address cumulative impacts
  - *Hammond v. Norton* (D.D.C.)

10.10 EA case law and EAs

Because EISs generally are more publicized documents addressing larger or more controversial projects, the case law involving EISs is greater than that specifically addressing EAs. Nevertheless, the number of EAs prepared each year is substantially larger than the number of EISs, and agencies have increasingly turned to EAs to expedite NEPA compliance for actions with questionable levels of impact. Swartz preformed a review of substantive NEPA cases issued by federal courts in 2005.[74] The following five cases all considered the extent to which public involvement requirements applied to EAs.

Sierra Nevada Forest Protection Campaign v. Weingardt (E.D. Cal. 2005): In this case, the district court held that while the CEQ regulations "do not require the circulation of a draft EA, they do require that the public be given as much environmental information as is practicable, prior to completion of the EA, so that the public has a sufficient basis to address those subject areas that the agency must consider in preparing the EA." The agency's scoping notices were insufficient because they did not contain an analysis of the environmental impacts of the projects. The need for effective communication with the public in EAs as well as EISs cannot be overemphasized.

Alliance to Protect Nantucket Sound, Inc. v. U.S. Department of the Army (1st Cir. 2005): The plaintiffs argued that the Army Corps should have issued a draft EA for public comment. However, the U.S. Court of Appeals held that the agency met the "to the extent practicable" requirement by issuing public notice of the proponent's application, providing a comment period that was later extended to more than five months, carrying out two public hearings, noting and responding to public comments in the EA, and conferring with federal and state environmental agencies. The court noted that nothing "in the CEQ regulations requires circulation of a draft EA for public comment, except under certain 'limited circumstances.' 40 C.F.R. §1501.4(e)(2)."

The reader should note that the court confused the requirement for public involvement in preparing EAs in 1501.4(b) with the requirement to issue a FONSI in draft in special circumstances in §1501.4(e)(2).

El Dorado County v. Norton (E.D. Cal. 2005): In this case, the district court noted that the CEQ regulation requiring public involvement in EAs to the fullest extent practicable has been interpreted "to mean that 'the public must be given an opportunity to comment on draft EAs and EISs.'"[75] The agencies did issue the draft casino EA for public review and were not required to circulate the FONSI.

TOMAC v. Norton (D.D.C. 2005): Here, the district court held that the agency met the requirements for public involvement for its supplemental EA and FONSI. The court concluded that there had been extensive opportunities for formal and informal public comment in this case. The court noted "FONSIs must be made available for public review, 40 C.F.R. § 1501.4(e)(1), as was done here, but there is no explicit statutory or regulatory requirement that EAs be submitted for public comment."

Natural Resources Defense Council v. U.S. Forest Service No. 04-35868, 35 ELR 20160 (9th Cir. Aug. 5, 2005): In this case, district court held that the Forest Service violated NEPA by failing to provide for effective predecisional public involvement in the preparation of the EAs for four logging projects. The agency countered that issuing a scoping notice and releasing the final EA to the public satisfied the mandatory public involvement requirements. The court noted that while:

> the CEQ regulations do not require the circulation of a draft EA, they do require that the public be given as much environmental information as is practicable, prior to completion of the EA, so that the public has a sufficient basis to address those subject areas that the agency must consider in preparing the EA.

The scoping notices contained no analysis of the impacts of the projects and failed to give the public adequate information to effectively participate in the decision making process. The court enjoined the projects until the Forest Service complied with NEPA.

10.11 Study of alternatives analysis in NEPA documents

Smith recently completed a study of challenges to alternatives analyses in NEPA documents.[76] These challenges involved U.S. Federal Courts of Appeals cases over the period 1996–2005. The study showed that agencies were predominantly successful in defending against challenges to their alternatives analyses. In this study, federal agencies won 30 of the 37 cases. The study identified five lessons and some practical steps to prepare an alternatives analysis. These cases' decisions indicate that where a federal agency prepares an appropriate statement of purpose and need, analyzes in detail the range of reasonable alternatives that meet the stated purpose and need, and provide rationale for dismissing other alternatives from detailed analysis, it will generally be successful if it faces future litigation on the analysis of alternatives. Although some of the lessons are drawn from cases specifically addressing the alternatives analysis in EISs, all of the lessons are applicable to the consideration of alternatives in EAs as well.

Lesson 1: Agencies should explain their rational for determining the range of reasonable alternatives analyzed. This study found that in virtually every case that challenged an agency's exclusion of an alternative from detailed analysis, the agency prevailed if it had explained its reasoning and lost if it were silent. The most legally sound reason is that the alternative does not meet the stated purpose and need for agency action. Many of the court decisions noted that there is no set minimum number of alternatives required to be analyzed in an NEPA document. The principal driver for the range of necessary alternatives is reasonableness, not number.

Lesson 2: Agencies should carefully consider a request from another agency, individual, or organization to consider an alternative in detail because it is deemed reasonable. Courts have made it clear that agencies are not obligated to fully analyze every alternative suggested to them, but if a suggested alternative can be construed as "reasonable," it must be analyzed unless a well-reasoned explanation is provided. In all seven cases lost by federal agencies, they did not adequately explain their decisions not to fully analyze suggested additional alternatives.

Lesson 3: Agencies should explain their statement of purpose and need. Courts generally deferred to an agency's statement, except where the agency did not provide a valid reason for constructing its statement as it did, or narrowed the statement so much that only one alternative was reasonable. In several cases, the courts indicated that federal agencies preparing NEPA analyses supporting decisions to grant permits or approvals to private-party applicants could give substantial weight to the purpose and need of the applicants. In other words, the needs of applicants can be a permissible justification for an agency to narrow its range of reasonable alternatives.

Lesson 4: Inadequacies in consideration of the no-action alternative does not appear to be a major legal vulnerability. Agencies were challenged on construction of the no-action alternative five times during the 10-year period, and prevailed in all five cases. Every EIS or EA should always, however, contain a careful consideration of the no-action alternative.

Lesson 5: Analysis of only two alternatives in an EA may be appropriate under certain circumstances. In the three cases that addressed this issue, the courts found in favor of the agency and noted that where analysis of the proposed action unquestionably revealed no potential for significant impacts, developing additional alternatives was not required by NEPA or the CEQ regulations. One decision noted that several courts have agreed that "the obligation to consider alternatives in an EA is a lesser one than under an EIS." Lesser does not imply nonexistent. In general, a two alternative approach is more likely to be acceptable for short EAs for simple actions than for longer EAs for more complex or controversial actions.

10.12 Other cases reflecting on the preparation of EAs

Other recent cases involving EAs provide additional guidance for preparing defensible EAs.

10.12.1 Avoid implying that a FONSI is predetermined

In one case, plaintiffs used e-mails in the administrative record and obtained through discovery legal evidence of the federal agencies' inappropriate determination to issue a FONSI before completing the environmental analysis. The agencies would have been prudent to avoid assuming a FONSI early in the project, e.g., by qualifying such phrases with "unless significant impacts are identified."

10.12.2 Participate actively as cooperating agency

It can be a risky position for a cooperating agency to remain passive during the EA process. In one case, the DOE relied on the lead agency to scrutinize the EA sections on toxic air emissions and regulatory requirements. Later, when this EA air emission analysis became the subject of litigation, it became clear that as a potential codefendant a cooperating agency (in this case the DOE) cannot afford such reliance but must itself review the internal draft NEPA document carefully.

References

1. Swartz, L., *Recent Nepa Cases*, National Association of Environmental Professionals 35th annual conference, Session #5, Atlanta GA, April 27–30, 2010.
2. Dickerson, R., *The Interpretation and Application of Statutes*, Little, Brown & Co., Boston, 1975.
3. Black, H.C., *Black's Law Dictionary*, abridged 5th ed. West Publishing, St. Paul, MN, 1983.
4. *Sierra Club v. Peterson*, 717 F.2d 1409 (D.C. Cir. 1983).
5. Boggs, J.P., Procedural vs. Substantive in NEPA Law: Cutting the Gordian Knot, *The Environmental Professional*, Vol. 15, pp. 25–34, 1993.
6. Black, H.C., *Black's Law Dictionary*, 6th ed., West Publishing, St. Paul, MN, 1990.
7. *Calvert Cliffs'Coordinating Comm. v. AEC*, 449 F.2d 1109 (U.S.C.A, D.C. Cir. 1971).
8. *Kleppe v. Sierra Club*, 427 U.S. 390 (1976).
9. *Vermont Yankee Nuclear Power Corp. v. Natural Resources Defense Fund, Inc.*, 435 U.S. 519(1978).
10. *Strycker's Bay Neighborhood Council, Inc. v. Karlen*, 444 U.S. 223 (1980).
11. *Baltimore Gas & Electric Co. v. Natural Resources Defense Fund, Inc.*, 462 U.S. 87 (1983).
12. Fogleman, V.M., *Guide to the National Environmental Policy Act*, in citing *Robertson v. Methow Valley Citizens Council*.
13. Administrative Procedure Act, 5 U.S.C. 706.
14. CEQ, "Council on Environmental Quality—Forty Most Asked Questions Concerning CEQ's National Environmental Policy Act Regulations" (40 CFR 1500-1508), *Federal Register*, Vol. 46, No. 55, pp. 18026–38, March 23, 1981, question 12c.
15. Yost, N.C. and Rubin, J.W., The National Environmental Policy Act, unpublished, adapted from Yost, N.C., Administrative Implementation of and Judicial Review Under the National Environmental Policy Act, in *Law of Environmental Protection*, S. Novick, S., ed., Clark Boardman Company Ltd., New York, 1987.
16. Bell, C., Brownell, F.W., and Cardwell, R., *Environmental Law Handbook*, 12th ed., Chapter 10, Government Institutes, Lanham, MD. 1993.
17. Fogleman, Guide to the National Environmental Policy.
18. Bell, C., Brownell, F.W., and Cardwell, R., *Environmental Law Handbook*. 12th ed. Government Institutes, Lanham, MD. 1993.
19. Mandelker, D., *NEPA Law and Litigation*, 2d ed. Clark Boardman Callaghan. New York, NY. 1992.
20. *Calvert Cliffs'Coordinating Comm. v. AEC*, 449 F.2d 1109 (U.S.C.A., D.C. Cir. 1971).
21. *Citizens to Preserve Overton Park, Inc. v. Volpe*, 401 U.S. 402 (1971).

22. Mandelker, D.R., *NEPA Law and Litigation*, Section 4.04, Clark Boardman and Callaghan, New York, NY. 1993.
23. *Citizens to Preserve Overton Park, Inc. v. Volpe*, 401 U.S. 402 (1971).
24. *Cronin v. Dept. of Agriculture*, 919 F.2d 439 (7th Cir. 1990).
25. *Hughes River Watershed Conservancy v. Glickman*, 81 F.3d 437 (4th Cir. 1996).
26. *Haynes v. United States*, 891 F.2d 235 (9th Cir. 1989).
27. *County of Suffolk v. Secretary of the Interior*, 562 F.2d 1368 (2nd Cir. 1977).
28. *Cronin v. Dept. of Agriculture*, 919 F.2d 439 (7th Cir. 1990).
29. *County of Suffolk v. Secretary of the Interior*, 562 F.2d 1368 (2nd Cir. 1977).
30. *Animal Defense Council v. Hodel*, 840 F.2d 1432 (9th Cir. 1988).
31. *Asarco, Inc. v. EPA*, 616 F.2d 1153 (9th Cir. 1980).
32. *Asarco, Inc. v. EPA*, 616 F.2d 1153 (9th Cir. 1980).
33. *County of Suffolk v. Secretary of the Interior*, 562 F.2d 1368 (2nd Cir. 1977).
34. Federal Rules of Evidence, Rule 702.
35. Federal Rules of Evidence, Rule 803(18).
36. Federal Rules of Evidence, Rule 803(8)(c).
37. *Citizens to Preserve Overton Park, Inc. v. Volpe*, 401 U.S. 402 (1971).
38. *Environmental Law Handbook*, Chapter 1.
39. 5 U.S.C. 706(2).
40. *Foundation on Economic Trends v. Lyng*, 680 F. Supp. 10 (D.D.C. 1988).
41. *Marsh v. Oregon Natural Resources Council*, 490 U.S. 360 (1989).
42. *Environmental Law Handbook*, Chapter 10.
43. *Citizens to Preserve Overton Park, Inc. v. Volpe*, 401 U.S. 402 (1971).
44. *Citizens to Preserve Overton Park, Inc. v. Volpe*, 401 U.S. 402 (1971).
45. 5 U.S.C. 706(2); also *March v. Oregon Natural Resources Council*, 490 U.S. 360 (1989).
46. *Marsh v. Oregon Natural Resources Council*, 490 U.S. 360 (1989).
47. *Marsh v. Oregon Natural Resources Council*, 490 U.S. 360 (1989).
48. *Sierra Club v. U.S. Army Corps of Engineers*, 772 F.2d 1043 (2d Cir. 1985).
49. *Marsh v. Oregon Natural Resources Council*, 490 U.S. 360 (1989).
50. *Marsh v. Oregon Natural Resources Council*, 490 U.S. 360 (1989).
51. *Marsh v. Oregon Natural Resources Council*, 490 U.S. 360 (1989).
52. Mandelker, D.R., *Guide to the National Environmental Policy Act*, 2d ed., Clark Boardman and Callaghan, 1996.
53. Rodgers, *Handbook on Environmental Law*, Section 7.4.
54. Eccleston, C.H., NEPA: Determining When an Analysis Contains Sufficient Detail to Provide Adequate NEPA Coverage for a Proposed Action, *Federal Facilities Environmental Journals*, Number 2, pp. 37–50, Summer 1995.
55. Mandelker, *NEPA Law and Litigation*, Section 3.04.
56. 541 F.2d 1018, 1038 (4th Cir. 1976).
57. *Citizens to Preserve Overton Park, Inc. v. Volpe*, 401 U.S. 402 (1971).
58. *Citizens to Preserve Overton Park, Inc. v. Volpe*, 401 U.S. 402 (1971).
59. Rodgers, *Handbook on Environmental Law*, Chapter 7.
60. Rodgers, *Handbook on Environmental Law*, Chapter 7.
61. Rodgers, *Handbook on Environmental Law*, Chapter 7.
62. Rodgers, *Handbook on Environmental Law*, Section 7.1
63. Rodgers, *Handbook on Environmental Law*, 731–3.
64. *Half Moon Bay Fisherman's Marketing Association v. Carlucci*, 857 F. Supp. 505, 512 (U.S. App., 9th Circ. 1988).

65. *Blue Ocean Preservation Society v. Watklns,* 754 F. Supp. 1450, 1459 (1991); *Puna Speaks v. Edwards,* 554 F. Supp.
66. *Holy Cross Wilderness Fund v. Madigan,* 960 F.2d 1515, 1522 (1992).
67. *Sierra Club v. Marsh,* 976 F. Supp. 763, 770 (1992).
68. *City of Tenakee Springs v. Clough,* 915 F.2d 1308, 1312 (1990).
69. *Sierra Club v. Pen/old,* 664 F. Supp. 1299, 1314 (1987).
70. *Robertson v. Methow Valley Citizen Council,* 490 U.S. 332, 335 (1989).
71. 28 U.S.C. §2412 (b) and (d), 1988.
72. *United States v. Rainbow Family,* 695 F. Supp. 314, 325 (E.D. Tex. 1988).
73. Swartz, L., *Recent NEPA Cases,* 2005, http://www.nepa.gov/nepa/caselaw/NEPA_Cases_2005_NAEP_paper.pdfhttp://www.nepa.gov/nepa/caselaw/NEPA_Cases_2005_NAEP_paper.pdf.
74. Swartz, L., *Recent NEPA Cases,* 2005, http://www.nepa.gov/nepa/caselaw/NEPA_Cases_2005_NAEP_paper.pdfhttp://www.nepa.gov/nepa/caselaw/NEPA_Cases_2005_NAEP_paper.pdf.
75. Citizens for Better Forestry, 341 F.3d at 970 (quoting *Anderson v. Evans,* 314 F.3d 1006, 1016 (9th Cir. 2002)).
76. Smith, M., US DOE NEPA Lessons Learned, Is This Reasonable? *A Review of NEPA Alternatives Analysis Case Law,* September 4, 2007, Issue #52, p. 16.

chapter eleven

Specialized non-NEPA environmental assessment documents

11.1 Introduction

While the NEPA environmental assessment (EA) is the focus of this book, there are also a number of environmental investigations that are closely related and sometimes even casually but erroneously referred to as "environmental assessments." A detailed discussion of the history, preparation, use, and interpretation of these related environmental assessment documents, at a level presented in the preceding chapters for the NEPA EA, is outside the scope of this book. However, the preparer or user of NEPA EAs should be aware of these related environmental documents. Moreover, many analytical principles outlined in the previous chapters for NEPA EAs apply equally to these other documents; in fact, many of the related documents are often prepared simultaneously on the same projects for which NEPA EAs are developed and may, in fact, be referenced in NEPA EAs.

The following sections briefly introduce some of these more common and related environmental assessment documents.

11.2 Phase I environmental site assessment

The 1980 Comprehensive Environmental Response, Compensation, and Liability Act (CERCLA)[1], better known as "Superfund," is considered by many to be one of the most important pieces of environmental legislation enacted in the United States. NEPA is widely known only among government workers, environmental contractors, lawyers, and environmental activists; "Superfund," by contrast, is almost become a household word. CERCLA is a product of the same environmental movement in the 1960s and 1970s that engendered NEPA, and probably would never have developed in its present form had it not been for NEPA, but it is perhaps a more widely recognized statute than NEPA. Interestingly, it has been stated that NEPA provided the model on which much of the CERCLA was based upon.

The environmental regulations of the 1960s and 1970s, including (among others) NEPA, Clean Water Act, and Clean Air Act, are directed at preventing future pollution and other environmental degradation. CERCLA instead largely addresses environmental remediation of contamination that has already occurred. Just as a number of headline incidents in the 1960s such as the Santa Barbara oil spill and Cuyahoga River fires helped spur enactment of NEPA, these same incidents combined with similar incidents hitting the headlines in the 1970s helped spur CERCLA. Chief among the latter were the contamination discovered in Love Canal in New York and Times Beach, Missouri. The Times Beach incident was particularly poignant—a small town not unlike towns habited by many Americans had to be completely abandoned and the residents relocated. The subsequent rapid discovery of other contamination sites engendered widespread fear among large segments of the American population.

While responsibility for achieving the goals of NEPA was placed directly on the federal government, and responsibility for achieving the goals of prescriptive environmental statutes such as the Clean Air Act and Clean Water Act was usually placed on industry, responsibility for achieving the environmental cleanups mandated by CERCLA could not always be so easily assigned. Many contaminated sites were no longer owned by the party that perpetuated the pollution. In many instances, the party responsible for the contamination was a company or corporation that had gone bankrupt or otherwise been legally dissolved. The property had often been sold to other business interests innocent and unaware of the contamination on the real estate they acquired. Use of the terms "compensation" and "liability" in the name of the act reflects this challenge. Whenever a "potentially responsible party" (PRP) could be identified, it could be required to foot at least part of the cleanup cost. However, the act required the federal government to pay for cleanups where no PRP could be identified. To do so, the government established a trust fund colloquially referred to as the "Superfund." It is from this trust fund that the act gained its informal but commonly referenced name.

While a handful of large and severely contaminated sites prompted the initial enactment of Superfund, its effects rapidly reached into the nooks and crannies of America, especially in the older industrial regions of the Northeast and Midwest. Following a course of action outlined in a national contingency plan for cleanups under Superfund[2], the Environmental Protection Agency (EPA) tasked a series of field investigation teams (FITs) to investigate thousands of potentially contaminated sites. The teams visited the sites, reviewed file records, and interviewed people familiar with the history of the site, a process termed a preliminary assessment (PA). If the PA revealed possible contamination, the teams collected a limited number of representative samples of environmental media such as soil, surface water, and groundwater for chemical analysis, a process termed a site investigation (SI). The PA and SI for many sites were performed concurrently and presented in a single preliminary assessment/site investigation or PA/SI report. The PA/SI process served to screen out sites lacking contamination requiring cleanup. The remaining shortlist of sites was then subjected to a more extensive program of environmental sampling, termed a remedial investigation (RI), to characterize and delineate areas of contamination. Sometimes multiple rounds of sampling and analysis were needed to complete an RI.

Once the RI was complete, attention turned to selecting a cleanup method. The process used to choose a cleanup method, termed a feasibility study (FS), is surprisingly similar to the NEPA process. Like an EIS or EA, an FS evaluates multiple actions to accomplishing a cleanup, and considers the possible effects of not taking any action (a "no-action" alternative). An FS evaluates the effects and efficacy of each cleanup alternative and also considers a key factor not evaluated in an EIS—cost. The selected cleanup alternative must be effective and technologically feasible, and it must accomplish the cleanup objectives at a reasonable cost. Sometimes the RI and FS are reported together in a single remedial investigation/feasibility study (RI/FS) report. Once a design alternative is selected, design and implementation can follow.

One particularly controversial element of CERCLA was that property owners could be held responsible for cleanup costs for contamination that occurred before they purchased the property. Only if property owners could prove that they had no reasonable way of knowing the property was contaminated prior to purchase could they clearly escape liability. The procedure for claiming this "innocent landowner defense" is contained in Section 120(h) of CERCLA. It is not simple: The property purchaser must pursue "due diligence" to investigate the history and records for the property for evidence of possible contamination. The purchaser is not expected to do an exhaustive investigation of every conceivable information source that might provide evidence of possible prior contamination, but they are expected to make a reasonable effort to track down and examine those sources that can be accessed without excessive cost or delay.

The need for a due diligence investigation of the environmental history of a property prior to purchase became cemented once banks loaning money to finance the purchases realized that purchasers burdened by environmental cleanup costs could default on their loans. Banks started requiring due diligence environmental surveys prior to lending money to purchase commercial property. Soon, thousands of such surveys were being conducted every year throughout the United States. And unlike NEPA EISs and EAs, these surveys were required by the private sector, of corporate America.

A need was immediately identified for detailed guidance on just what constituted an adequate survey. Terms like "due diligence" are vague at best. Banks, businesses, and individuals needed precise step-by-step instructions on how to conduct a legally defensible due diligence investigation that effectively eliminated the risk of being held responsible for cleanup costs but that did not substantially increase the costs of property acquisition. The American Society for Testing and Materials (ASTM; currently ASTM International) stepped up to the plate and issued a standard in 1993 termed "Standard for Conducting Phase I Environmental Site Assessments" and designated ASTM Standard E 1527-93[3]. Unlike the Council on Environmental Quality (CEQ), ASTM is not a government or even quasi-government entity. Founded in 1898, it consists of a private membership that develops, approves, and updates a series of technical standards and practices for application in multiple fields of science and engineering. The process for developing a standard is different but no less rigorous than the federal rulemaking process (Administrative Procedures Act). Subcommittees of members draft a standard that is then distributed by committees to their larger membership for formal review and voting. Members may vote to accept the standard, reject it, or accept it with changes. Large or controversial standards can go through multiple rounds of balloting before receiving enough votes for acceptance. The process sometimes takes longer than two years. Written suggestions by voters are evaluated for possible inclusion in the standard following each round of balloting. Each standard thereby represents a professional consensus.

ASTM Standard E 1527-93 was revised through formal ASTM balloting in 1995, 1997, 2000, and 2005. The latest version is ASTM Standard E 1527-05[4]. It is henceforth referred to in the following text as "ASTM 1527." Although differing in certain details, each version called for performing the following activities:

- A visual site inspection, involving walking the property looking for visible evidence of possible past contamination.
- A records review, involving review of government and private records for evidence of possible past contamination.
- Interviews with current and former occupants of the property.

Examples of items commonly looked for during the visual site inspection include drums, soil stains, dead or dying vegetation or wildlife, ground disturbances possibly indicative of landfill activity, and foundations of former buildings that may have been contamination sources. Examples of records typically reviewed include government databases of spills, inspections, and other environmental incidents; file records maintained by current and past site occupants; maps and building blueprints showing the locations for past activities; and photographs and aerial photographs depicting past conditions at the property. Because of the vast range of records that could theoretically be tracked down and reviewed, ASTM narrowed the field of records constituting due diligence to those that were:

- Reasonably ascertainable, i.e., readily obtained through minimal efforts within a short time.
- Practicably reviewable, i.e., readily interpreted by a broadly knowledgeable environmental professional.

ASTM 1527 also calls for performing a title search of owners of record for the property. Owner names suggestive of industrial or waste management operations capable of releasing substantial quantities of environmental contamination could provide evidence of possible past contamination.

ASTM 1527 somewhat parallels the process followed by EPA and its contractors when performing PAs. It does not call for sampling environmental media, as would be performed in an SI. Sampling, if desired, would be deferred to a Phase II environment site assessment (ESA) following procedures in another ASTM standard, ASTM E 1903-97[5]. Whereas EPA usually seeks to further reduce uncertainty through a modest sampling effort before concluding that no cleanup is needed, and hence usually follows a PA with an SI, a bank or prospective purchaser is more prone to walking away from a property if an initial review indicated enough uncertainty to warrant sampling. Hence, many more Phase I ESAs are performed than Phase II ESAs.

Although NEPA EAs and Phase I ESAs are both relatively short environmental documents, they actually share few similarities. NEPA EAs look forward to what the possible effects of an action will be on the environment, while a Phase I ESA looks back at the history of possible environmental contamination. NEPA EAs must consider the full range of environmental resources and impacts, while Phase I ESAs are focused on contamination, safety, and health concerns. NEPA EAs evaluate alternatives, while Phase I ESAs are baseline documents only, with no actions or alternatives. An NEPA EA supports a finding of no significant impact (FONSI) or indicates a need to prepare an EIS, while a Phase I ESA supports a decision to either purchase or not purchase a property.

In one respect, a Phase I ESA is like an NEPA EA: It is a short analytical effort that can be used to determine whether a larger analytical effort, the Phase II ESA, is needed. In this line of reasoning, a Phase I ESA is somewhat analogous to an EA and a Phase II ESA is somewhat analogous to an EIS. But a Phase II ESA does not closely resemble an EIS; like a Phase I ESA, it is exclusively a baseline document that does not evaluate a contemplated future action or alternatives to such an action. Interestingly, ASTM has developed a standard for a very brief screening document, termed a transaction screen (ASTM E 1528)[6], that can be used to evaluate whether the effort of a Phase I ESA is needed. Transaction screens cover similar data sources and lines of reasoning as Phase I ESAs but consist of simple checklists that can be completed in only a few hours. In this way, the transaction screen might be considered as somewhat analogous to an EA, at least the very short EAs initially envisioned by CEQ, while the Phase I ESA might be considered somewhat analogous to an EIS.

Phase I ESAs are generally short documents. Unlike NEPA documents, the need for Phase I ESAs is not limited to the federal government. Hence, considerably more Phase I ESAs are prepared annually than are NEPA documents. A relatively limited cadre of environmental contractors operate in the NEPA market, while thousands of individuals and businesses, especially small businesses, operate in the Phase I ESA market. From the early to mid-2000, an average price for preparing a Phase I ESA generally ran under $5,000, while the price for an EA could easily top $50,000. Most Phase I ESA preparers work from boilerplate report structures developed following a recommended outline included in ASTM E 1527. While most NEPA EAs also follow a common basic outline, they are generally much more customized reports.

Like NEPA EAs, Phase I ESAs, are often prepared by broadly trained environmental generalists as well as specialized experts. Whereas the need for specialized expertise in NEPA is usually defined by the nature of the proposed action and anticipated environmental impacts, specialized expertise in waste management, chemistry, and biology can be highly useful for Phase I ESAs. Expertise in geology, hydrology, and hydrogeology can be particularly useful, considering the need to understand fate and transport of chemical contamination in soils, water, and other media.

The scope of Phase I ESAs expanded to include many other environmental topics following implementation of the 2000 version of ASTM 1527 (ASTM E 1527-00)[7]. This version encouraged Phase I ESA preparers to address topics related to "business environmental risk" in addition to the traditional waste management and contamination issues considered previously. "Business environmental risk" includes the presence of features potentially limiting economic use of the property such as wetlands, floodplains, historical and archaeological resources, and threatened or endangered species. Banks recognized that the presence of these features could limit economic development of the property and thereby affect the probability of defaulted loans. A Phase I ESA addressing "business environmental risk" as well as traditional environmental due diligence issues thereby provided a more complete assessment of the economic risk posed by environmental issues. The range of topics addressed by Phase I ESAs therefore became much closer to that addressed in NEPA EAs and EISs.

The potential consequences of errors and oversights in a Phase I ESA can be even greater than in an NEPA document. Property purchases could be saddled with millions of dollars in cleanup costs or litigation fees. Recognizing this situation, EPA implemented more stringent educational and experience requirements for Phase I ESA preparers, staring in 2006, than the earlier general guidelines established by ASTM. Phase I ESA preparers now must possess a professional license in engineering or a related state license plus three years of relevant experience, a bachelor's or higher degree in a relevant field of science or engineering and five years of relevant experience, or ten years of relevant experience. Preparers must state and certify by signature that they meet these requirements in each Phase I ESA they perform. However, there are still no nationwide requirements for formal certification of Phase I ESA preparers.

NEPA EA preparers may find that Phase I ESAs are very useful sources of information if available for properties that would be occupied by the proposed action or an alternative. Phase I ESAs contain baseline information on the environmental setting of a property such as its geology, hydrogeology, topography, climate, soils, and (especially if they address "business environmental risk") ecology and cultural history. This information can be used in writing the affected environment section of the NEPA EA. Conversely, Phase I ESA preparers can find NEPA EAs and EISs to be useful environmental records with information on past land use and waste management practices on properties currently or formerly owned by the federal government.

Additionally, services exist that offer rapid, inexpensive database review services to serve Phase I ESA preparers. They offer a database review package covering most standard environmental record sources specified in ASTM E 1527 for less than $2,000. The packages could traditionally be obtained in five to seven days and in less than three days if a rush fee were paid. In the last few years, customers could access the packages instantaneously online if they first established accounts and passwords to track billing. Furthermore, the environmental database services recognized that the information they provide can be of use to customers preparing NEPA documents as well as Phase I ESAs. Some now offer "NEPA packages" with prepackaged baseline data covering information services such as National Wetland Inventory Maps (for rough locations of wetlands), Flood Insurance Rate Maps (for floodplains), Natural Heritage Maps (for threatened and endangered species), and Cultural Resource Maps (for archaeological and historic resources). The NEPA packages can greatly expedite writing the affected environment chapter of an NEPA EA or EIS but, of course, do not contribute directly to the impact analysis.

11.3 Cell tower NEPA review

The 1990s, especially from 1995 to 1999, witnessed the construction of a vast new infrastructure of telecommunication receivers needed to operate cellular telephones (cell phones). Each major provider of cell phone services needed to install thousands of receivers on tall structures. The receivers could sometimes be mounted on the rooftops of tall buildings in urban areas but in rural and most suburban areas usually had to be mounted on towers. The tall steel lattice poles and tall single pole towers (monopoles) gradually became familiar features on the American landscape, especially in metropolitan areas and near interstate highways.

Each receiver had to be permitted by the Federal Communications Commission (FCC), a federal action subject to NEPA[8]. The FCC needed a way to determine whether the action could have a potential for significant impacts on the human environment and therefore require an EIS. Receiver installation posed the potential for several categories of potentially significant environmental impacts, especially with respect to natural and cultural resources. Hence, blanket coverage under a categorical exclusion (CATEX) was not possible. However, the effort needed to prepare formal EAs for each receiver was daunting. So, the FCC developed a simple initial procedure for screening proposed receiver installation sites[9]. Sites passing the screen would need no further NEPA review (could be categorically excluded). Other sites would need further analysis, at a minimum an EA[10].

The screening procedure involves evaluating areas potentially affected by a cell tower for the presence of the following:

1. Officially designated wilderness areas.
2. Officially designated wildlife preserves.
3. Officially designated critical habitats, or areas having threatened or endangered species that are likely to be jeopardized.
4. Districts, sites, buildings, structures, or objects significant in American history, architecture, archaeology, engineering, or culture that are listed, or eligible for listing, in the National Register of Historic Places.
5. American Indian religious sites.
6. Floodplains.
7. Areas that would require significant change in surface features such as wetlands, streams, or forests.
8. Use of high density lighting in residential areas.
9. Excessive radio frequency radiation levels in certain areas.

The screening reports were generally only two or three pages plus supporting documentation such as federal wetland and floodplain maps and State Historic Preservation Office (SHPO) letters regarding potential historic and cultural resource impacts. Consultants sometimes reported the screens in short letter reports. At most a few sentences of text would be provided to document each of the criteria. In many respects, the reports fit the CEQ's original intent for an EA: a concise analytical document to ascertain whether further analysis is necessary. The criteria generally cover most of the environmental issues typically covered in an EA, with the notable exception of socioeconomic issues such as housing and employment.

Chapter eleven: Specialized non-NEPA environmental documents 261

Ironically, the expectation was that sites failing the screen would next proceed to an EA. However, very few did. Most sites failing the screen were abandoned. Communications companies developing the towers generated large groups of candidate sites and could usually jettison a few individual sites not passing the screen. That fact, is not, however, negative nor in contradiction to NEPA's objectives; in fact making efficient and timely decisions regarding the siting of projects and avoidance of adverse environmental effects without the need for long and costly analyses is an exemplary application of NEPA.

Despite the similarities, the FCC screening reports differed from EAs in several ways. They did not evaluate alternatives (and they did not evaluate a "no-action" alternative). They did not describe purpose and need. They did not include a FONSI. Perhaps of most concern they did not consider cumulative impacts. Each cell tower was evaluated individually without consideration of the fact that most towers were parts of proposals to cluster numerous towers in closely spaced locations over regional landscapes. Failure to consider cumulative impacts generally did not endanger discretely positioned features such as wilderness areas, wetlands, or floodplains that could be clearly avoided. But while the distant visibility of one cell tower several miles out from a historic feature could be, and often was, considered not to fail the screen based on the cultural resource screening factor. Nevertheless, the distant visibility of numerous towers might in actuality be a potentially significant impact worthy of an EA or even EIS.

A rush of cell tower screens was performed by many environmental consulting firms in the late 1990s. Contracts would commonly call for addressing sites in packages of 20, 50, or more sites in a locality. Consultants would visit up to a dozen or more sites per day taking notes and photographs, write or visit regulatory offices to collect background data, and prepare templated reports. Bidding competition became fierce; companies began looking for package bids where the cost per site was less than $2,000. The NEPA reviews were sometimes packaged with Phase I ESAs for the sites; developers could expect bids of $2,500 to $5,000 per site to provide both environmental services.

Companies that maintained environmental databases began to offer expedited review packages. The availability of computerized FCC NEPA review packages further reduced the bids relative to the older individualized packages. Once the intensive wave of cell tower development neared its conclusion (a lower rate of tower development continues still), these companies began to market the FCC NEPA review packages as generalized sources of some of the most common baseline data included in the affected environment sections of EAs and EISs. The packages could not, and cannot, provide all of the baseline data needed for EISs or most major EAs, but they could substantially reduce the effort needed for baseline data collection for many smaller EAs. They could also reduce the effort needed to collect some of the simplest routine baseline data used in EISs and larger EAs, such as floodplain map data and National Wetland Inventory map data. The packages obviated the need to visit agency offices to collect the maps. However, the more recent posting of mapped data as geographic information system (GIS) layers posted on agency websites has made it easy for EA and EIS preparers to collect the data directly without paying for search packages.

11.4   Biological assessment

The biological assessment (BA) is to the Endangered Species Act of 1972 (ESA)* as the EIS is to NEPA: It is the analytical document that supports decisions made and actions taken to comply with the ESA. The ESA is one of a flurry of environmental statutes that, along with NEPA, was introduced as part of the environmental movement in the early 1970s. Perhaps no environmental statute has been as controversial as the ESA. Many unfamiliar with NEPA have heard of the snail darter and northern spotted owl, two species protected under the ESA that have slowed or hindered key economic development projects. Whereas NEPA and most provisions of the Clean Air Act and Clean Water Act were not substantially changed over the years, implementation of the ESA was substantially altered to make the statute more palatable to private landowners. Many prescriptive limitations on land use were adjusted in favor of more flexible habitat conservation plans that attempted to reconcile species protection with economic objectives. Most recently, the U.S. Fish & Wildlife Service (FWS) on behalf of the outgoing George W. Bush administration implemented a "midnight rule" relaxing interagency consultation requirements under the act[11]. The rule was officially revoked in April 2009.

The ESA targets protection of species designated by the federal government as endangered or threatened. Endangered species are defined under the act as those "in danger of extinction throughout all or a significant portion of its range." Threatened species are defined under the act as those "likely to become an endangered species within the foreseeable future throughout all or a significant part of its range." Even though the ESA places greatest urgency on endangered species, practitioners under the act commonly refer to the protected species as "threatened or endangered species." Species are placed on the list, and hence offered protection under the act only after considerable deliberation. Species are initially proposed for listing by the FWS or the National Marine Fisheries Service (NMFS) following collection of adequate supporting data. These

---

* The reader should note that as used in this context (i.e., BA) the acronym "ESA" should not be confused with the environmental site assessment, which was discussed earlier.

two agencies are collectively referred to in the context of the ESA as the services, with the FWS taking the lead on terrestrial and inland aquatic species and the NMFS taking the lead on coastal aquatic species. Before a species proposed for listing can actually be listed, the proposed listing must be announced in the *Federal Register* and the public be given an opportunity to comment. Similar announcements must be made to remove a species from the list or change the designation of a listed species from threatened to endangered, or vice versa.

The expectation is that once a species is listed and afforded protection under the ESA, its population will recover to the point that continued protection is unnecessary. The service (FWS or NMFS) proposing a listing concurrently prepares a recovery plan outlining metrics for ascertaining whether the species has recovered to the point that removal from the list (termed "delisting") is justified. The most famous delisting under the ESA is the bald eagle, whose population numbers had become so depleted by the widespread use of the agricultural insecticide DDT that it was listed as endangered. Populations increased steadily since widespread DDT use ceased in the 1970s, so much so, that the FWS was able to upgrade its status to threatened in 1995 and actually delist it in 2005. Delisting the bald eagle was hailed in 2005 as evidence of the success of the ESA.

There are two principal regulatory components to the ESA. The first is contained in Section 9 of the act, which prohibits anyone from "taking" a listed species without a permit, termed a "take permit." The term "take" is defined under the act as harassing, harming, pursuing, hunting, shooting, wounding, killing, trapping, capturing, or collecting the species. The prohibitions contained in Section 9 apply to all, whether in the federal, state and local, or private sectors. The notion that species in peril of extinction should not be hunted or killed is not generally controversial even amongst most critics of the act. Whether "harm" extends to habitat degradation, as might result from development of privately owned land containing habitat, is highly controversial.

The second regulatory component is contained in Section 7 of the act, which requires federal agencies to consult with the services (FWS and NMFS) when conducting, authorizing, or funding an activity with the potential to adversely affect listed species. Section 7 therefore applies to most federal actions requiring an EIS or EA. It can also apply to actions categorically excluded under NEPA if those actions have any potential to adversely affect listed species. Although the term "consult" suggests that mere notification or an informal discussion might suffice, as described in the following paragraphs, the consultation process required under the ESA is rigorously defined.

First, the responsible agency must determine whether its proposed action could adversely affect a listed species. The agency or its consultants usually begins by requesting data on known locations of, or potential areas of occurrence for, listed species. After determining whether any listed species could potentially occur in the affected area, agencies typically engage in "informal" consultation. The agency submits to the services a letter describing the proposed action and a map of the affected area. The services respond with a letter indicating that no listed species likely occur in the area, thereby concluding the ESA consultation process, or that listed species might occur. If the services think that listed species could occur in the affected area, they will request that the agency submit a BA[12].

BAs can be thought of as focused impact assessment documents for listed species. They generally read much like an EA or EIS, beginning with a description of the proposed action and description of baseline conditions followed by a description of possible impacts and mitigation measures. One key difference with NEPA documents is that BAs do not have to address alternatives, even the "no-action" alternative. Environmental considerations in the federal decision to proceed with an action or select an alternative falls under the purview of NEPA; the BA serves only to assess whether the services need to formally review the action and recommend measures to reduce or mitigate impacts. If, as a result of an EA or EIS, an agency selects a different alternative following completion of a BA for a previously contemplated action, it must, of course, prepare a new BA for the new alternative.

Agencies may prepare BAs that exceed minimum requirements under the ESA. For example, the Forest Service prepares "biological evaluations" that address species considered to be regionally sensitive as well as federally listed species. Figure 11.1 is an outline for a BA for a non-native invasive plant control project on the Ottawa National Forest in Michigan. Note that the BA considers alternatives as well as a proposed action. That approach proved advantageous for the Forest Service; based on comments received on the accompanying draft EA, the Forest Service ultimately selected an alternative involving a broader scope of treatment and did not have to revise the biological evaluation (BE). Note that the BE provides separate chapters addressing federally listed species (the true BA) and nonlisted species identified by the Forest Service as regional forester sensitive species (RSFF).

A BA makes one of three possible determinations for each species it addresses:

- No effect.
- May affect, is not likely to adversely affect.
- May affect, and is likely to adversely affect.

The "no effect" determination is similar to an NEPA determination of "no significant impact." The "may affect, is not likely to adversely affect" determination is used when the impacts are expected to be adverse but

1.0 INTRODUCTION
2.0 LOCATION OF PROJECT
3.0 DESCRIPTION OF THE ALTERNATIVES
    3.1 PURPOSE AND NEED
    3.2 PROJECT DESCRIPTION
        3.2.1 Alternative 1: No Action
        3.2.2 Alternative 2: Integrated NNIP Control Program (Proposed Action)
        3.2.3 Alternative 3: Increased Extent of Treatment
        3.2.4 No Use of Biological Controls
4.0 LISTING OF SENSITIVE SPECIES
5.0 SENSITIVE SPECIES DESCRIPTION AND IMPACT ASSESSMENT
6.0 RECOMMENDATIONS
7.0 RISK ASSESSMENT
8.0 DETERMINATIONS
9.0 MONITORING
10.0 REFERENCES

Figure 11.1 Biological evaluation outline, Ottawa National Forest Non-Native Invasive Plant Control Program.

minimal or expected to be beneficial. It is commonly used in situations where practices such as clearing and grubbing might benefit species favoring open rather than forested landscapes. The "is likely to adversely effect" determination is similar to an NEPA determination of potentially significant impacts. Note, however, that the determinations are made for each species, not each action. A single action might have determinations of "no effect" for some species, "is not likely to adversely affect" for some species, and "is likely to adversely affect" for some species.

If a BA includes any "is likely to adversely affect" determinations, the responsible agency must formally consult with the FWS (for terrestrial or inland aquatic species) or NMFS (for coastal aquatic species). In a formal consultation, the services issue a biological opinion (BO) describing the potential effects and recommending any reasonable and prudent alternatives that might be possible to reduce or offset (mitigate) the effects. The agency is required to consider but not necessarily implement the alternatives. With respect to indicating a need for formal consultation, a BA functions much like an EA: both indicate that no further action is necessary if they demonstrate no potentially significant adverse impacts. However, the technical scope of a BA is much more similar to the biological sections of an EIS than an EA. It provides the services with technical data for use in preparing a BO should formal consultation be required. No subsequent technical documentation (analogous to an EIS) is prepared. The BO more closely resembles an NEPA Record of Decision (ROD) than an EIS.

Traditionally, BAs were prepared as separate documents, even for projects addressed by an EA or EIS. Of course, the biological resources sections of the EA or EIS for a project addressed in a BA usually reported much, if not most, of the same data and analysis included in the BA. The best EAs and EISs only summarized the BA: limitations on the length and complexity of text in NEPA documents do not apply to BAs, which unlike NEPA documents are technical scientific documents intended primarily for use by other technical experts. One change recently implemented in January 2009 under the outgoing Bush administration ("midnight rule") changed the ESA (but subsequently revoked in April 2009), permitting agencies to merely provide the EA or EIS to the services rather than a separate stand-alone BA document. This change might have made sense as long as the EIS or EA provided all of the technical detail that would have otherwise been included in the separate BA. If, however, the rule changes resulted in excessively lengthy or complicated NEPA documents, or worse, allowed agencies to fail to provide the services with the technical detail needed to support BO preparation, they would have constituted a setback for both NEPA and the ESA. It remains to be seen whether the rule changes will be reimplemented and, if so, constitute a step forward or backward for environmental protection.

11.5 Environmental assessments and state NEPA programs

According to the CEQ, twenty states and other United States jurisdictions such as Guam and the District of Columbia have implemented state regulations that at least partially parallel the federal NEPA process[13]. These state "mini-NEPAs" have the same general objectives as the federal NEPA. Most use the term "environmental policy" in the regulations and most require an interdisciplinary evaluation of potential environmental impacts for state agency actions with a potential to significantly affect the environment. Unlike the federal NEPA, some states limit their NEPA-like reviews to actions directly undertaken by state agencies. Others extend their reviews to permitting decisions by state agencies, thereby assessing private development actions requiring state permits. State processes for preparing and reviewing environmental impact documents vary widely, and not all use the federal NEPA terminology, including words such as EIS, EA, or CATEX. However, most have a process for documenting environmental impacts that generally resembles the EIS, a process for screening actions for potentially significant environmental impacts that generally resembles the EA, and a process for exempting small and routine actions that generally resembles the CATEX.

As part of a book focusing on the EA, the following discussion of state NEPA processes focuses on state processes that screen actions for potentially significant environmental impacts—what can be thought of as the "state EA." Most NEPA processes rely more on checklists and standardized forms and less on text to accomplish the impact screening function of the federal EA. The use of checklists or forms reflects a general simplicity of documentation inherent in many state processes relative to the federal NEPA process. However, most state processes require a brief formal statement similar to a FONSI that posits that no potentially significant environmental impacts would result from the subject action and that no EIS is therefore necessary. Unlike the federal EA, many states such as New York require completion of an EA-like checklist even if an EIS is contemplated. The EA-like analysis therefore forms a necessary preliminary step for an EIS.

The biggest difference between the federal and state processes is that the EA-like processes for most states do not require evaluation of alternatives (most of the EIS-like processes do require documentation of alternatives). The logic of evaluating alternatives in federal EAs for actions lacking potentially significant environmental impacts can be questioned. Detailed documentation of alternatives in federal EAs is generally limited to the so-called "long" EAs that have become common practice only in recent years and were never contemplated at the time the CEQ guidelines were issued. Alternatives evaluation in the "short" EAs originally contemplated by CEQ is generally perfunctory. Most state EA-like processes, and especially the state checklists, more closely resemble the original "short" EA than the "long" EA. Although many states call for EA-like equivalent documents or checklists without formal evaluation of alternatives, the act of preparing the documents or checklists still provides an opportunity for agencies to stop and think whether minor modifications might be available to reduce environmental impacts.

As a general rule, the most populous states with densely populated urban regions tend to have environmental review processes for state projects that most resemble NEPA. One state process quite similar to the federal NEPA process is the California Environmental Quality Act (CEQA) process. The basic environmental impact assessment document under CEQA is the environmental impact report (EIR), which is analogous to an EIS. If an agency can document that an action not exempt from CEQA would have no significant environmental impacts, it can prepare a negative declaration, which is analogous to a FONSI. A negative declaration is defined as "a written statement briefly describing the reasons that a proposed project will not have a significant effect on the environment and does not require the preparation of an environmental impact report."[14] CEQA states that negative declarations must include:

- A brief description of the project, including a commonly used name for the project, if any.
- The location of the project, preferably shown on a map, and the name of the project proponent.
- A proposed finding that the project will not have a significant effect on the environment.
- An attached copy of the initial study documenting reasons to support the finding.
- Mitigation measures, if any, included in the project to avoid potentially significant effects.

The initial study is analogous to an EA. Like a federal EA/FONSI, negative declarations and their associated initial studies must be circulated for public review. As for the federal EA, the CEQA regulations do not prescribe a specific format for the initial study. As for an EA, it simply must contain enough information to adequately support the conclusion of no significant impact in the negative declaration. In a parallel to the mitigated FONSI, "mitigated negative declarations" can be prepared for projects with significant impacts if the proponent can commit to mitigation that effectively offsets the impacts.

Another highly populous and urban state with a process highly similar to NEPA is New York. The basic environmental impact assessment document under the New York State Environmental Quality Review Act (SEQR) is the state EIS. As in California, if an agency can document that an action not exempt from SEQR would have no significant environmental impacts, it can prepare a negative declaration. Similar to California and to the federal mitigated FONSI, New York agencies can issue "conditioned negative declarations" for projects relying on mitigation to avoid potentially significant environmental impacts.

The technical review document underlying the New York negative declaration is the environmental assessment form (EAF). Similar to the federal CATEX process, the state has identified categories of "Type II" actions not requiring an EAF or other further action under SEQR. Other categories of actions are referred to as "Type I" and must be analyzed using the EAF. The technical scope of the EAF generally mirrors the range of environmental issues commonly considered in federal EAs. Most questions in the EAF request yes/no or single-word responses but provide space for explanatory text. The checklist structure of the EAF discourages lengthy text on resources clearly not impacted by, or issues clearly not relevant to, the subject action but allows for an expanded text-based

Chapter eleven: Specialized non-NEPA environmental documents   269

discussion where detailed analysis is needed. The EAF does not require (or provide space for) information on alternatives.

In a process with no clearly defined federal NEPA counterpart, the state has identified "listed" Type I actions requiring analysis using the full EAF. Other Type I actions are assumed to be "unlisted" and may be analyzed using an abbreviated version of the EAF. The state emphasizes that use of the abbreviated EAF does not necessary lead to a decision to issue a negative declaration. "Listed" actions are categories of actions widely recognized as having a known substantial risk for significant adverse environmental impacts, while the risk for "unlisted" is less clear. The use of full and abbreviated EAFs somewhat parallels the increasing use by many federal agencies of "long" EAs for complex projects instead of the traditional 10- to 15-page EAs encouraged by CEQ.

New Jersey's Executive Order 215 of 1989[15] establishes a state NEPA-like program applicable only to projects directly initiated by state agencies or that are at least 20 percent state funded. It uses the federal EIS and EA nomenclature from NEPA, but uses a radically different basis for determining which document is required. The EA is used for "Level 1" projects, defined as those whose cost exceeds one million dollars; and the EIS is used for "Level 2" projects, defined as those whose cost exceeds five million dollars and whose land disturbance exceeds 5 acres. Projects not meeting the "Level 1" or "Level 2" criteria are excluded from the order in a type of simplistic CATEX. Cost is not a consideration in evaluating whether federal projects require an EA or EIS, and extent of land disturbance is but one of multiple factors that must be simultaneously considered in determining whether a project qualifies for a FONSI.

Guidance prepared by the New Jersey Department of Environmental Protection (NJDEP)[16] states that state EAs are less comprehensive and less rigorous documents than state EISs. Furthermore, it states that state EAs do not require evaluation of project alternatives, while state EISs do. However, the New Jersey EA is a text document, not a checklist. The non-evaluation of alternatives is the chief difference between New Jersey and federal EAs. Consistent with the federal CEQ guidance, the NJDEP states that state EISs and EAs should be prepared "using a systematic interdisciplinary approach that will insure the integrated use of the natural and social sciences and the environmental design arts." The guidance indicates that the level of project description and graphics in an EA should be comparable to those in an EIS but that the description of the existing environment and level of impact analysis should be briefer. The NJDEP guidance also provides more detailed direction for analyzing impacts to specific resources than does the federal CEQ.

Maryland is one of several states to have a "mini-NEPA" that calls for an NEPA-like review of projects directly carried out by state agencies. But what sets Maryland apart is a law enacted in 1991 that requires an environmental baseline document for private land development projects. The law is the Maryland Forest Conservation Act[17], modeled somewhat after tree protection ordinances previously passed by certain local jurisdictions in Maryland. The act requires developers to prepare two-phased documents. The first, the forest stand delineation (FSD) serves to inventory baseline conditions. The second, the forest conservation plan (FCP) serves to identify and quantify resource losses[18].

Although as expected the FSD and FCP must address trees and forest cover, the act requires a more comprehensive approach to inventorying and preserving surface landscape features. Surface features that must be considered in FSDs and FCPs include, among others:

- Forest cover.
- Specimen trees, individual trees exceeding specified size thresholds.
- Streams and state-regulated stream buffers.
- Wetlands and state-regulated wetland buffers.
- Floodplains.
- Slopes exceeding specified grades.
- Soils.
- Prime farmland.
- Federal and state threatened or endangered species.
- Historic and archaeological features.

With the exception of socioeconomic resources, the requirements closely parallel the resource types addressed in many EAs and EISs. The FSD closely parallels the affected environment section and the FCP closely parallels the environmental consequences and mitigation sections. The FSD and FCP differ from an EA or EIS chiefly in two ways: first, the FSD and FCP do not address alternatives; and second, the FSD and FCP do not consider cumulative impacts. The developer is expected to use the FSD data to minimize resources impacts and therefore must think of alternative approaches for fitting the construction footprint on a site. However, the developer is not required to evaluate alternative sites for the development. The developer is required to evaluate how site work can affect resources on adjacent sites but is not required to consider cumulative impacts to the regional landscape.

In an interesting parallel to NEPA, the state initially required developers to prepare complete FSDs and FCPs for any land development (other than those in two sparsely populated counties and projects on sites smaller than 40,000 square feet)[19]. The legislature subsequently amended the act to allow for "simplified" or "intermediate" FSDs and FCPs for certain development projects on sites with no or minimal forest cover. The "simplified" documents can be considered roughly parallel to a CATEX, the "intermediate" documents can be considered roughly parallel to an EA, and the complete documents can be considered to be parallel to an

EIS. Like NEPA, the objective of the act is not excellent FSDs and FCPs but excellent protection of the resources they cover.

While the net accomplishment of many agencies' NEPA processes in protecting environmental resources is open to debate, the Forest Conservation Act has a clear record as an action-forcing mechanism. The state has estimated that approximately 79,174 acres of forest cover have been retained (preserved) on development sites as a result of the first ten years of implementation of the act (1992–2002). The state has also estimated that the act has resulted in planting approximately 13,611 acres of new forest cover over the same period[20]. The state does not encourage preparation of lengthy FSD or FCP texts, even allowing simple FSDs and FCPs to be presented solely on drawings in a manner similar to simple erosion and sediment control plans. Standard worksheets are used to present most data. Much of the FCP focuses on mitigation identification of areas of existing forest cover for preservation within development projects, specification of measures to protect preserved trees and forest cover during grading activities, and proposal of compensatory forest cover planting projects.

As is true in most states, developers of private land development projects in Maryland are not required to prepare EAs or EISs. But the Forest Conservation Act compliance process involves inventorying and evaluating impacts to many of the same resources considered in NEPA. It is unlikely that the Maryland legislature could ever succeed in passing legislation subjecting private developers to a requirement to prepare NEPA-like documents. But the Maryland Forest Conservation Act has helped introduce NEPA-like natural resource planning objectives into private sector land development activities.

References

1. 42 U.S.C. § 9601 et seq.
2. 40 C.F.R. Part 300.
3. American Society for Testing and Materials (ASTM) E 1527-93, Standard Practice for Environmental Site Assessments: Phase I Environmental Site Assessment Process, prepared by ASTM Subcommittee E50.02 on Commercial Real Estate Transactions under ASTM Committee E50 on Environmental Assessment.
4. American Society for Testing and Materials (ASTM) E 1527-05, Standard Practice for Environmental Site Assessments: Phase I Environmental Site Assessment Process, prepared by ASTM Subcommittee E50.02 on Commercial Real Estate Transactions under ASTM Committee E50 on Environmental Assessment.
5. American Society for Testing and Materials (ASTM) E 1903-97 (Reapproved 2002), Standard Guide for Environmental Site Assessments: Phase II Environmental Site Assessment Process, prepared by ASTM Subcommittee E50.02 on Commercial Real Estate Transactions under ASTM Committee E50 on Environmental Assessment.

6. American Society for Testing and Materials (ASTM) E 1528-06, *Standard Practice for Limited Environmental Due Diligence: Transaction Screen Process*, prepared by ASTM Subcommittee E50.02 on Commercial Real Estate Transactions under ASTM Committee E50 on Environmental Assessment.
7. American Society for Testing and Materials (ASTM) E 1527-00, *Standard Practice for Environmental Site Assessments: Phase I Environmental Site Assessment Process*, prepared by ASTM Subcommittee E50.02 on Commercial Real Estate Transactions under ASTM Committee E50 on Environmental Assessment.
8. 47 C.F.R. Part 1.1301 et seq.
9. 47 C.F.R. Part 1.1307.
10. 47 C.F.R. Part 1.1308.
11. *Draft Environmental Assessment for the Proposed Modifications to Regulations Implementing Interagency Cooperation under the Endangered Species Act*, U.S. Fish and Wildlife Service, Department of the Interior and National Marine Fisheries Service, Department of Commerce, October 2008.
12. U.S. Fish and Wildlife Service and National Marine Fisheries Service, 1998, *Endangered Species Consultation Handbook*, Procedures for Conducting Consultation and Conference Activities under Section 7 of the Endangered Species Act, March.
13. CEQ, *State Environmental Planning Information*, updated March 19, 2009, available at http://ceq.hss.doe.gov/nepa/regs/states/states.cfm.
14. CEQA Section 21064.
15. New Jersey Executive Order No. 215 of 1989, *Environmental Assessment*.
16. Attachment to Executive Order No. 215 of 1989, "Guidelines for the Preparation of an Environmental Impact Statement/Environmental Assessment," revised and updated April 23, 2002, New Jersey Department of Environmental Protection, Trenton, New Jersey.
17. *Natural Resources* Article 5-1601 et seq., Annotated Code of Maryland.
18. Maryland Department of Natural Resources, 1997, *State Forest Conservation Technical Manual*, 3rd ed., Ginger Page Howell and Tod Ericson, eds.
19. Maryland Department of Natural Resources, December 31, 1991, *Forest Conservation Manual: Guidance for the Conservation of Maryland's Forests During Land Use Changes*, under the 1991 Forest Conservation Act.
20. Maryland Department of Natural Resources, September 2004, *The Maryland Forest Conservation Act: A Ten Year Review*.

chapter twelve

Summary

> Whenever you find yourself on the side of the majority, it is time to pause and reflect.
>
> —**Mark Twain**

12.1 Epilogue

The objective of this book has been to provide a unique perspective on the National Environmental Policy Act (NEPA) focusing on the environmental assessment (EA) rather than the better known and more intensively scrutinized environmental impact statement (EIS). Given that most agencies prepare such a disproportionately greater number of EAs than EISs, and that the trend over the past several years has been an ever increasing reliance on EAs rather than EISs, agencies are clearly going to great lengths to avoid preparing EISs. This is partially a result of agencies seeking an acceptable way around the more rigorous documentation and public involvement requirements and associated expenses of an EIS. However, in a more positive light, the increased reliance on EAs can also be partly attributed to the fact that agencies are learning to better integrate NEPA objectives (e.g., proposing lower impact alternatives, integrating mitigation measures into project designs) into early planning, thereby reducing potential impacts before actions mature to the documentation stage.

As emphasized throughout the book, an EA and finding of no significant impact (FONSI) can usually be more difficult to defend than an EIS and Record of Decision (ROD), because the burden of proof is on the agency to demonstrate that significant impacts would not result or that any such impacts can be mitigated. An EIS does not carry this burden of proof. For this reason, adversaries have increasingly focused efforts on challenging EAs, which are considered more vulnerable targets; agencies can defend against this strategy by devoting particular care to the preparation of their EAs and in exercising prudence in reaching a conclusion regarding significance. Alternatively, adversaries may increase their odds of success by directing efforts at flawed or poorly prepared EAs and FONSIs. The need for good science and clear presentation are just as great for an EA as for an EIS, perhaps even more so. The EA and FONSI process, while always intended to provide simplify the documentation process for actions clearly lacking a potential for significant environmental impacts, were never intended as a mechanism for skirting the need for a hard look at—and public scrutiny of—actions that could have significant environmental impacts.

Conclusions regarding significance are reserved for the decision maker, based on the information presented in the EA. The decision maker is responsible for ensuring that a FONSI is not issued for an action where an EIS is warranted.

FONSIs that make cursory, sweeping, or unsubstantiated conclusions are at particular risk of a successful challenge. While the process of preparing an EA and FONSI must include opportunities for public involvement, it is a less publicly visible process than an EIS with fewer opportunities for public input. The Council on Environmental Quality (CEQ) intended for the EA and FONSI process to serve as a simplified and streamlined process to expedite NEPA review for projects lacking significant environmental impacts; it was never intended to be a process for greater concealment of actions that might actually have potentially significant environmental impacts. CEQ has cautioned that repeated failure to involve the public can constitute a violation of the NEPA regulations, providing adversaries with sufficient grounds for a successful legal challenge.

An EA must provide a rigorous, scientifically accurate, and unbiased basis for assessing potential significance. The analysis should be prepared so as to clearly assist the reader in drawing conclusions regarding significance. All investigated environmental impacts should be clearly documented and based on professionally accepted scientific methodologies. The quality of scientific analysis and clarity of writing in an EA must be comparable to that in an EIS. Brevity can be achieved in an EA by focusing discussion on environmental issues necessary to support a FONSI and address potential uncertainty or controversy regarding the FONSI, but not by a perfunctory analysis. Decisions based on an EIS and FONSI must be as scientifically defensible as those based on an EA.

With respect to the regulatory vacuum that exists in providing detailed direction on preparing EAs, this text strives to help bridge this void. By applying the "rule of reason," it identifies relevant EIS regulatory requirements that can also be logically interpreted to apply to preparation of EAs. The text has also attempted to capture the professional experience and best professional practices from seasoned practitioners who have spent years preparing EAs for multiple agencies. Lacking detailed regulatory direction, the focus has been on providing the reader with a reasonable, definitive, consistent, and comprehensive methodology for managing, analyzing, and writing EAs.

Index

Page numbers followed by f indicate figure
Page numbers followed by t indicate table

A

Accident analyses
  accident consequences, 111–112
  accident scenarios and probabilities, 109–111
  intentionally destructive acts, 112
  overview, 108–109
  risk and, 111–113
  sliding scale, 108, 108t
Accident consequences
  involved and noninvolved workers, 111–112
  uncertainty and, 112
Action-impact model, 42f
Action(s)
  alternative, 155
  component, 155
  concept of, 52
  connected, 100, 102
  human environment and, 54
  interim, 30, 31t
  needs addressed in EA, 158–159, 158t
  potential, 141
  proposal, 200
  proposed, 71–72, 77, 162, 163
  significance of, 53
  site-specific, 53, 196
Activity-related requirements, 187t
Activity test, 180t, 181, 183, 188t
Acts, intentionally destructive, 112
Adequacy
  characteristics of, 242–243, 243t
  Mandelker definition of, 241
Adequate administrative record, 234

Administrative law. *see* Code of Federal Regulations (CFR)
Administrative Procedures Act, 227, 236–237, 256
Administrative record
  case law and, 78–79
  reviewing, 234–236
ADREC. *see* Agency administrative record (ADREC)
Adverse impacts, 55t, 93, 94
AEC. *see* U.S. Atomic Energy Commission (AEC)
Affected environment
  defining region of influence, 160
  interpreting baseline measurements, 159–160, 160t
  need for, 160–161
Affecting, 54–55
Agency administrative record (ADREC), 207, 208t
Agency(ies). *see also* Federal agencies
  bringing suits against, 231–232
  categorical exclusion and, 56
  cooperating with, 75
  final decision, 241–242
  listing of, 173–175
  reviewing administrative record, 234–236
  state, 139
  strategy of, 233
Agricultural Stabilization and Conservation Administration, 5
Air quality standards, conformity determinations with, 169
Alliance to Protect Nantucket Sound, Inc. v. U.S. Department of the Army (1st Cir. 2005), 247
Alternative analysis, NEPA documents and, 248–249

Alternatives
  comparing, 90, 90t, 164–165
  comparing impacts among, 171–172
  defining range of, 88–89
  identifying, 62f, 64–65, 65t, 85–86
  intensity, 86
  no-action, 91, 157, 162–163, 162t
  numbers in the analysis, 88
  range of, 86–88
  reasonable, 163–164, 164t
  recommendations for, 92t
  route, 86
  schedule, 86
  site, 86
  technology, 86
American Association for the Advancement of Science, 4
American Council of Governmental Industrial Hygienists, 188
American Society for Testing and Materials (ASTM). *see* ASTM International
Andrews Air Force Base, Maryland, 163
Antiquities Act in 1906, 5
APA. *see* U.S. Administrative Procedures Act (APA)
Approvals, listing of, 172–173
"Arbitrary and capricious" standard, 238, 239–240
"Arbitrary and capricious" test, 233–234, 237
Ashley Creek Phosphate Co. v. Norton (9th Cir.), 246
Assessing significance, 127
  definitions and use of, 177–178
  introduction to, 177
  need for systematic approach, 179
  potential significance, 178–179
  procedure for, 179–197, 180t
  significant beneficial impacts, 197–200
Assumptions, documenting, 140–141
ASTM International, 256
ASTM Standard E 1527-93, 256–257

B

BA. *see* Biological assessment (BA)
Baltimore Gas and Electric Co. (BG&E), 52, 226
Baseline measurements, interpreting, 159–160, 160t
Beneficial environmental impacts, 90t, 197, 198, 204, 205
"Best management practices" (BMPs), 171

BG&E. *see* Baltimore Gas and Electric Co. (BG&E)
Bidding, 135–137, 149, 261
Biological assessment (BA), 262–266
Biological evaluation (BE), 264–265, 265f
Biological impacts, 95–97, 96t
Biological opinion (BO), 265
Blogs, 140
BMPs. *see* "Best management practices" (BMPs)
BO. *see* Biological opinion (BO)
Bounding analysis, 93–94
Bounding approach, 110, 112
Building and Construction Trades Department of the AFL-CIO, 12
Bureau of Reclamation, 6, 7

C

Caldwell, Lynton, 7
California Environmental Quality Act (CEQA), 267
Carson, Rachel, 5–6
Carter, Jimmy, 11, 12
Case law
  administrative record and, 78–79
  categorical exclusion and, 23–24
  impact analysis and, 168
  involving NEPA, 244–245
Categorical exclusion (CATEX), 3
  agencies and, 56
  case law and, 23–24
  documenting, 57–58
  guidance on applying, 57
  as a level of NEPA compliance, 22
  NEPA review and, 121
CATEX. *see* Categorical exclusion (CATEX)
Cause-and-effect relationships, evaluating, 99
Cell tower NEPA review, 259–262
CEQ. *see* Council on Environmental Quality (CEQ)
CEQA. *see* California Environmental Quality Act (CEQA)
"CEQ NEPA regulations." *see* regulations
CERCLA. *see* Superfund
CFR. *see* Code of Federal Regulations (CFR)
Chafee, John, 13
City of Oxford v. Federal Aviation Administration (11th Cir.), 246
City of Riverview v. Surface Transportation Board (6th Cir.), 246

Index

City of Shoreacres v. Waterworth (5th Cir.), 246
Civilian Conservation Corps, 5
Class action suits, 231–232
Clean Air Act, 1955, 1, 6, 14, 20, 254, 262
Clean Air conformity requirements, 113–114
Clean Water Act, 3, 14, 20, 115, 168, 173, 214, 221, 254, 262
Clean water Permitting Programs, 168
Cliffs, Calvert, 11
Clinton, Bill, 119, 143
Coastal Zone Management Act-Federal Consistency, 117
Code of Federal Regulations (CFR), 16, 222–223
Comparing alternatives, 90, 90t, 164–165
Compensation, and Liability Act (CERCLA) "Superfund." *see* Superfund
Component actions, identifying, 155
Comprehensive Environmental Response, 20, 54, 141
Conformity determination process, 114
Conformity determinations with air quality standards, 169
Conformity review process, 114
Connected actions, 100, 102
Conservatisms, 111
Consultation, 68–69
Contents of environmental assessment, 148–150, 149t
Context, 53, 54t, 196
Contractors, environmental assessment and, 69–70, 134, 149
Contracts, types of, 134–135
Controversial, 195
Controversy test, 180t, 181, 195, 195t
Cost reimbursable contracts, 134–135
Costs
 direct and indirect, 135
 legal, recovering, 244
Council on Economic Advisers, 7, 20
Council on Environmental Quality (CEQ), 137, 256
 creation of, 7, 16
 on "detailed statement", 10
 documentation requirements for an EA, 149t
 draft regulations, 12
 environmental assessment and, 36
 guidance, 120–121, 123
 interpretation of "Major", 51
 on mitigation and monitoring, 127–130

NEPA implementing regulations, 11–12
 regulations of, 43, 44
 Title I and II, 19–20
Court's role
 agency's final decision and, 241–242
 EIS requirements and, 238–240
 focus of courts, 242
 supplementing EIS, 240–241
Court's rule, 233–234
Cultural practice, environmental impacts and, 125
Cumulative effects analysis, in EAs, 97
Cumulative environmental exposure, 126t
Cumulative impact analyses, 97–99, 102, 192
Cumulative impacts, 97, 101t, 126, 167, 191–192
Cumulative test, 180t, 181, 190, 193–194, 193t, 195t

D

Data Quality Act, 117
DBS. *see* Decision-based scoping (DBS)
Decision-based scoping (DBS), 70
Decision-Identification Tree (DIT), 70
Decision in principle, 194
Decision making, finding of no significant impacts and, 205–207, 206t
Decision process, attributes of, 65t
Decision Protocol 2.0, of U.S. Forest Service, 64–65
Decisions, words to be used in, 141
Declaratory judgment, 244
Depositions and reports, 80
The "detailed statement", 8–9, 10
Dingell, John, 7
Direct and indirect impacts, 166
Discovery, 231
Discovery process, 80
DIT. *see* Decision-Identification Tree (DIT)
Documentation requirements, 157–175. *see also* Environmental assessment documents
 affected environment, 159–161
 agencies and people consulted listing, 173–175
 environmental impact, 165–172
 finding of no significant impacts (FONSI), 209–211, 209t
 need for taking action, 158–159, 158t
 permits, approvals, and regulatory requirements listing, 172–173
 preparers listing, 175

proposed action and alternatives, 161–165, 162t
DOE. *see* U.S. Department of Energy (DOE)
Doub, Peyton, 236
Douglas County v. Babbit, 198–200
Due diligence, 255–256, 257, 258

E

EA. *see* Environmental assessment (EA)
EAF. *see* Environmental assessment form (EAF)
EA/FONSI documents, 46
Early and open process, 32–33, 34
Ecology Center v. Austin (9th Cir.), 246
Editing, EA documents and, 146, 148
Effects
    human health, 106–107, 107t
    significant, 178–179
EIR. *see* Environmental impact report (EIR)
EIS. *see* Environmental impact statement (EIS)
EJ. *see* Environmental justice (EJ)
El Dorado County v. Norton (E.D. Cal.), 246, 247
E-mails, administrative records and, 235
Endangered Species Act (ESA), 3, 4, 14, 57, 69, 118, 174, 198
Environment, categories of, 56
Environmental assessment analysis
    decision-based scoping, 70
    impacts analysis, 71
    proposed action, 71–72
    significance determinations for FONSI, 72
Environmental assessment compliance process
    cooperating with other agencies, 75
    reducing the length of EA, 73–75
    tiering, 76
    time and money, 75–76
Environmental assessment documents. *see also* Documentation requirements; Specialized non-NEPA environmental assessment documents
    editing, 146, 148
    figures and, 144
    FONSI and, 46
    glossaries and, 146
    graphic and visual aids and, 143–144, 145f
    Latin binomial names and, 147–148
    measurements units and use of, 147
    need section in, 158–159, 158t
    plain language and, 141–142, 143, 143t
    public, 67–69, 68t, 69t
    tables and, 144–145
Environmental assessment (EA). *see also* Environmental assessment analysis; Environmental assessment compliance process; Environmental assessment documents; Environmental assessment planning; Environmental assessment process; Environmental assessment *vs.* Environmental impact statement; Writing an environmental assessment (EA)
    analyses of, 36–37, 37t
    Council on Environmental Quality and, 36
    environmental justice and, 121
    as federal action, 21
    incomplete and unavailing information, 37–38
    long *vs.* short, 58–59, 59t
    NEPA compliance, 22
    preparation, 25, 134–147, 249–250
    purpose of, 45–46, 66, 66t
    state NEPA programs, 266–271
Environmental assessment form (EAF), 268–269
Environmental assessment planning
    applicants and contractors, 69–70
    consultation, 68–69
    EA are public documents, 67
    identifying alternatives, 62f, 64–65, 65t
    potential of significance and, 65–66. 62f
    public involvement, 62f, 63–64
    public notification, 67–68, 68t
    purposes served, 66, 66t
    scoping and public meetings, 69, 69t
    timing, 66
Environmental assessment process. *see also* Environmental assessment compliance process; Environmental assessment planning
    administrative record, 78–79
    introduction to, 61
    issuing FONSI, 77–78, 78t
    reducing the length of EA, 73–76
    serving as expert witness, 79–80, 81t

Index

Environmental assessment vs.
  environmental impact statement.
  see also Environmental
  assessment (EA); Environmental
  impact statement (EIS)
  actions, 52
  affecting, 54–55
  categorical exclusions, 56–58
  environmental impacts, 102, 104–105t, 105
  federal agencies, 51
  human environment, 55–56
  introduction to, 49–50
  legislation, 50
  major, 51
  proposals, 50
  public engagement and disclosure, 58
  significance, 52–54, 54t
Environmental consequences, 42f, 45
Environmental consulting services, 134–135
Environmental Defense Fund, 10
Environmental design arts, NEPA and integrating, 32
Environmental disturbances
  action and, 42f, 44
  identifying, 156, 157t
Environmental documents, 33–34, 152–153
Environmental effects, context and, 53, 54t
Environmental Effects Abroad: EO 12114, 116
Environmental generalist, 138
Environmental impact assessment (EIS). *see also* Environmental impacts
  accidental analyses, 107–110
  Clean air conformity requirements, 113–114
  environmental justice, 119–130
  floodplain and wetland environmental review requirements, 114–115
  human health effects, 106–107, 107t
  identifying alternatives, 85–91, 92t
  impacts analysis, 91, 93–106
  introduction to, 83
  National Historic Preservation Act and, 115–116
  sliding-scale approach and, 83–84
Environmental Impact Assessment Review, 144
Environmental impact report (EIR), 267
Environmental impacts, 122
  meaning and examples of, 156, 157t
  screen and analyze, 156–157

Environmental impact section, in environmental assessment document, 165–172
Environmental impact statement (EIS), 3, 12. *see also* Environmental assessment vs. environmental impact statement; Supplemental environmental impact statement (S-EIS)
  decisions and, 206t
  Muskie and, 8
  NEPA compliance and, 22
  NEPA's requirements for preparing, 26
  non-programmatic, 31t
  process of, 21
  programmatic, 31t
  reason for preparing, 27
  supplementing, 23, 24f, 27
Environmental Impact Statements: A Comprehensive Guide to Project and Strategic Planning, 78
Environmental Impact Statements (Eccleston), 42, 46, 97, 208
Environmental Justice: EO 12898, 116
Environmental justice (EJ)
  analyzing environmental impacts, 122
  assessing significance and mitigation measures, 127
  categorical exclusions, 121
  CEQ guidance, 120–121, 123
  definition of EPA, 119
  environmental assessments and, 121
  evaluating impacts, 122
  implementation of, 120
Environmental lawsuit requirements, 227–229
Environmental monitoring, 128–129
Environmental movement
  in early twentieth century, 5
  in nineteenth century, 4
  in 1960s, 5–6
Environmental receptor, 182
Environmental requirements
  Clean Air conformity requirements, 113–114
  executive order requirements, 116–117
  Floodplain and wetland environmental review requirements, 114–115
  National Historic Preservation Act, 115–116
  statutory requirements, 117–118
Environmental resources, 42f, 44–45

Environmental Response, Compensation, and Liability Act (CERCLA), 253
Environmental site assessment (phase I), 253–259
Environment brevity in EAs, need for, 160–161
EPA. *see* U.S. Environmental Protection Agency (EPA)
Equal Access of Justice Act, 244
ESA. *see* Endangered Species Act (ESA)
European Economic Community, 15
"Excess fatal cancers", 106
Executive orders
 Environmental Effects Abroad: EO 12114, 116
 Environmental Justice: EO 12898, 116
 Federal Actions to Address Environmental Justice in Minority Populations and Low-Income Populations (EO 12898), 119
 Floodplain Management executive order 11988, 114
 Invasive Species: EO 13112, 116–117
 Marine Protected Areas: EO 13158, 117
 New Jersey's Executive Order 215 of 1989, 269
 Protection of Wetlands executive order 11990, 114, 115
 requirements of, 116–117
Expert witness
 administrative records and, 236
 serving as, 79–80, 81t
Exposures, environmental impacts and, 125

F

Farmers Home Administration, 199
FCC. *see* Federal Communications Commission (FCC)
FCP. *see* Forest conservation plan (FCP)
Feasibility study (FS), 255
Federal Actions to Address Environmental Justice in Minority Populations and Low-Income Populations (Executive Order 12898), 119
Federal agencies
 categorical exclusions and the public, 58
 under NEPA, 51–52
 publishing and, 139
 using categorical exclusions, 57

Federal Communications Commission (FCC), 260, 261–262
Federal Emergency Management Agency, 154
Federalization, 52
Federalized agency, 51
Federal Register (FR), 15, 16, 222–223
Federal Rules of Civil Procedure, 231
Federal Water Pollution Control Act Amendments of 1972, 220
Figures, EA documents and, 144
Filing a suit, 231
Finding of No Significant Impact (FONSI)
 determining nonsignificance, 204–208
 determining significance, 72
 environmental impact statement and, 22
 introduction to, 203–204
 issuing, 77
 justifying for an action, 43
 mitigated, 211–215
 preparing, 208–211, 212t
 as public document, 77
Fish and Wildlife Coordination Act, 69, 174
Fixed price contracts, 134
Floodplain and wetland environmental review requirements, 114–115
Floodplain Management executive order 11988, 114
Floodplain or wetland assessment, for DOE, 115
FONSI. *See* Finding of No Significant Impact (FONSI)
Forest conservation plan (FCP), 270
Forest stand delineation (FSD), 270
FR. *see* Federal Register (FR)
Frequency scale, for risk analysis, 190t
FS. *see* Feasibility study (FS)
FSD. *see* Forest stand delineation (FSD)

G

General and administrative costs, 135
GIS. *see* Graphical information system (GIS)
Glacier National Park, 79
Global precedent, NEPA and, 14–15
Glossaries, EA documents and, 146
Graphical information system (GIS), 144
Graphic and visual aids, EA documents and, 143–144, 145f
Gray literature, 139
Greenhouse gas emissions, 53

Index

H

Hammond v. Norton (D.D.C.), 246
Hanly v. Kleindienst, 195
"Hard book" doctrine, 241
High and adverse impacts, evaluating, 122–123, 124–125t
Hiram Clarke Civic Club v. Lynn, 198
HUD. *see* U.S. Department of Housing and Urban Development (HUD)
Human environment, 54, 55–56
Human health effects, 106–107, 107t

I

Impacts. *see also* Cumulative impact; Cumulative impact analyses; Environmental impact assessment (EIS)
 adverse, 55t, 93, 94
 analysis of, 42f, 45, 71, 94–95, 168
 beneficial, 90t, 197, 198, 204, 205
 biological, 95–97, 96t
 bounding analysis, 93–94
 comparing among alternatives, 171–172
 describing, 167–168
 direct and indirect, 166
 in EA and EIS, 91, 93
 environmental, 122, 156, 157t
 evaluating, 122–123, 124–125t
 incomplete/unavailable information, 96–97
 on minority and low-income populations, 124
 recommendations, 103, 104–105t, 105
 reduction of, 211
 segmentation, 105–106
Inadequate analyses, characteristics of, 242–243, 243t
Information, dealing with, 37–38, 38t
Injunctions, 244–245
Injury in Fact test, 228
Intensity, 54, 55t
Intensity alternatives, 86
Intentionally destructive acts, 112
Interdisciplinary planning, 34–35
Interim actions, 30, 31t
Interim action status, eligibility for, 31–32
Invasive Species: EO 13112, 116–117
Irreparable injury, 227

J

Jackson, Henry (Scoop), 7, 8, 9, 10
Judicial review standards, 237
Jurisdiction, 227

L

Lake Erie, 6
Language, EA documents and, 141–142, 143, 143t
Latches, 227
Latin binomial names, EA documents and, 147–148
Legal costs, recovering, 244
Legal remedy, 234
Legal strategies, 232–233
Legal system, 230–2231
Legislation, 50
Legislative proposal, 50
Length, EA document and, 148
"Life's amenities", 18
Litigation process
 court's rule, 233–234
 enforcement of NEPA, 229–230
 legal remedy, 234
 legal strategies, 232–233
 legal system, 230–231
 reviewing an agency's administrative record, 234–235
 suing agencies, 231–232
"Localization" rule, 53

M

Magnuson-Stevens Fishery Conservation and Management Act-Essential Fish Habitat, 118
Major, 51
Man and Nature (Marsh), 4
Mandelker, D., 229, 241
Man made or built environs, 56
MAPs. *see* Mitigation Action Plans (MAPs)
March, Frederic, 177
Marine protected area (MPA), 117
Marine Protected Areas: EO 13158, 117
Marsh, George Perkins, 4, 5
Maximum concentration levels (MCLs), 186
MCLs. *see* Maximum concentration levels (MCLs)
Measurements units, EA documents and use of, 147

Methow Forest Watch v. U.S. Forest Service (D. Or), 246
Methow Valley, 226
Midnight rule, 262, 266
Mitigation, 42f, 46–47, 47t, 213
Mitigation Action Plans (MAPs), 46
Mitigation measures
  assessing, 127, 157
  CEQ and, 169–170
  criteria for adopting, 215t
  elements in, 170–171, 171t
  implementation by third party, 213
  implementing, 129–130
  ineffective, 128
  monitoring and documenting success of, 171
Monitoring, 42f, 47
  environmental, 128–129
  potential, assessing, 157
Mootness, 229
MPA. *see* Marine protected area (MPA)
Muir, John, 4, 5
Multidisciplinary approach, 34
Multiple environmental exposure, 126t
Murray, James, 6
Muskie, Edmund, 8

N

NAAQS. *see* National Ambient Air Quality Standards (NAAQS)
National Ambient Air Quality Standards (NAAQS), 113, 169
National Audubon Society, 10
National Emission Standards for Hazardous Air Pollutants, 186
National Employment Act of 1946, 7
National Environmental Policy Act of 1969 (NEPA). *see* NEPA
National Historic Preservation Act of 1969 (NHPA), 14, 57, 69, 115–116, 118, 174
National Marine Fisheries Service (NMFS), 129, 262
National Marine Sanctuaries Act, 118
National Register of Historic Places, 154
National Wetland Inventory maps, 154
Native Ecosystems Council v. U.S. Forest Service (9th Cir. August), 246
Natural and physical environs, 56
Natural Resources Conservation Service, 234
Natural Resources Defense Council, 10, 12
Natural Resources Defense Council v. U.S. Forest Service (9th Cir.), 246
Need section, in documents, 158–159, 158t
Negative declaration, 267–268
NEPA, 16, 173. *see also* Environmental movement; NEPA, legal standards and; NEPA and environmental impact analysis; NEPA brief overview; NEPA historical development; NEPA law and litigation; NEPA planning process; Regulations
  concept of scope, 34
  conducting an early and open process, 32–33, 34
  incorporating material by reference, 35–36
  integrating environmental design arts, 32
  integrating environmental planning, 32, 33t
  Interdisciplinary planning, 34–35
  interim actions, 30–32, 31t
  introduction to, 3
  judicial remedies in suits, 243–245
  as planning and decision making process, 30
  public involvement, 33–34
  requirements for preparing an EIS, 26
  systematic and interdisciplinary planning, 34–35
  writing documents in plain English, 35
NEPA, legal standards and
  Administrative Procedures Act, 236–237
  court's role, 238–243
  experts disagreement, 237–238
  judicial review standards, 237
NEPA and environmental impact analysis
  action-impact model, 42, 42f
  environmental disturbances, 44
  impact analysis, 45
  introduction to, 41–42
  mitigation, 46
  monitoring, 47
  receptors and resources, 44
  significance of, 45
NEPA and Environmental Planning (Eccleston), 21, 29, 31, 32, 34, 37, 45, 52, 70, 151, 158, 177, 211
NEPA and Environmental Planning Process (Eccleston), 50
NEPA brief overview
  mandate, 16

Index

procedure for passing laws and
regulations, 15–16
purpose of NEPA, 17, 30, 30t
regulatory nomenclature, 16
Title I, declaration of the National
Environmental Policy Act, 17
Title II, Council on Environmental
Quality, 17
NEPA historical development
birth of NEPA, 6–8
the "detailed statement," 8–9
effect around the world, 14
global precedent, 14–15
implementation, 10–11
implementation regulations, 11–12
passage of NEPA, 9–10
at present, 12–14
public comment review process, 12
NEPA law and litigation
alternative analysis in NEPA
documents, 248–249
applying legal standards, 236–243
case law involving NEPA, 245–246
EA case law and, 246–248
judicial remedies in NEPA suits,
243–244
lawsuit requirements, 227–229
legal interpretations of NEPA, 219–220
litigation process, 229–236
passing laws and regulations process,
220–223
preparation of EA cases, 249–250
regulatory requirements, 223–224
substantive *vs.* procedural process,
225–226
NEPA Law and Litigation (Mandelker), 220
NEPA planning process
beginning of, 21–22
completion of, 22
compliance levels, 22
environmental assessment, 25–26
environmental impact statement, 27
initiating NEPA process, 22–23, 24f, 25
requirements for preparing EIS, 26
NEPA Planning Process, 23, 53
NEPA's regulations, 20–21, 53, 223–224
NEPAssist, 155
New Jersey Department of Environmental
Protection (NJDEP), 269
New Jersey's Executive Order 215 of 1989,
269
New York State Environmental Quality
Review Act (SEQR), 268

1960s, environmental decade of, 5–6
Nineteenth century, environmental
movement in, 4
Nixon, Richard, 9, 10, 11, 12
NJDEP. *see* New Jersey Department of
Environmental Protection
(NJDEP)
NMFS. *see* National Marine Fisheries
Service (NMFS)
No-action alternatives, 91, 157, 162–163, 162t
Noise, corresponding context of, 54t
Nonlabor costs, 135
Non-programmatic EIS, 31t
NRC. *see* U.S. Nuclear Regulatory
Commission (NRC)
Nuclear Regulatory Commission (NRC), 11

O

Open Government Directive, 129
Organization, of environmental
assessment, 150–151
Organization for Economic Cooperation, 15
Ottawa National Forest in Michigan, 264
Outline, for environmental assessment, 149,
150, 150t
Overton Park case, 236, 238

P

PA. *see* Preliminary assessment (PA)
Pathways, environmental impacts and, 125
Peer-reviewed scientific journals, 139
People consulted list, in documentation,
173–175
Permanent injunction, 244
Permits, listing, 172–173
Phase I ESAs, 258–279
Photographs, EA documents and, 145–146
Photostimulations, EA documents and,
145–146
Pinchot, Giffort, 5
Plaintiff's strategy, 232–233
Postmonitoring, 47
Potential action, 141
Precedence test, 180t, 181, 194
Preliminary assessment (PA), 254
Preliminary Environmental Impact
Analysis, 182
Preliminary injunction, 244
Preparers list, in documentation, 175
Private lands, actions on, 100
Procedural process

NEPA and, 225–226
substantive *vs.*, 225
Programmatic EIS, 31t
Proposal action, 163, 200
Proposals
contracting and, 134–137
definition of, 50
Proposed action, 71–72, 77, 162–173, 163
Protection of Wetlands executive order 11990, 114, 115
Prudence, exercising, 232
Public document, FONSI as, 77
Public engagement and disclosure, 58
Public involvement
environmental assessment and, 62f, 63–64
NEPA and, 33–34
Public meetings, scoping and, 69
Public notification, 67–68, 68t

Q

Qualitative evaluation, cumulative impact analysis and, 100, 101t

R

RCRA. *see* Resource Conservation and Recovery Act (RCRA)
Readability, EA documents and, 141–142, 142t, 146
Reasonable, regulations and, 224
Reasonable alternatives, 163–164, 164t
Reasonably foreseeable, 107, 224
Receptor-related requirements, 184–186t
Receptors. *see also* Receptor test
checklist of, 182–183
environmental, 182
federal laws, regulations, and requirements, 184–186t
resources and, 44–45
Receptor test, 180t, 181
assessing, 182t
procedure for applying, 183, 187t
Record of Decision (ROD), 27, 141, 203–204, 222
Region of influence (ROI), 160
Regulatory compliance test, 180t, 181, 183, 188t
Regulatory guidance, 190
Regulatory requirements, listing, 172–173
Regulatory standards, impact analysis and, 168

Remedial investigation (RI), 255
Remedy(ies)
judicial, in NEPA suits, 243–244
legal, 234
types of, 244–245
Request for Proposals (RFPs), 135–137
Research, environmental assessment and, 139–140
Resource Conservation and Recovery Act (RCRA), 20
RFPs. *see* Request for Proposals (RFPs)
RI. *see* Remedial investigation (RI)
Ripeness, 227
Risk, accident and, 111
Risk analysis
frequency scale for, 190t
severity scale for, 190t
Risk/Uncertainty test, 180t, 181t, 189, 191t
Robertson v. Methow Valley Citizens Council, 226
ROD. *see* Record of Decision (ROD)
ROI. *see* Region of influence (ROI)
Roosevelt, Theodore, 5
Route alternative, 86
Rule of reason, 38, 194, 196
Rule of reason standard, 224

S

SAP. *see* Significant assessment test (SAP)
Schedule alternatives, 86
Scoping, 34
decision-based, 70
public meetings and, 69, 69t
Secondary information sources, 139
Segmentation, 50, 105–106, 155
SEQR. *see* New York State Environmental Quality Review Act (SEQR)
Severity-frequency determinations, significance assignments for, 191t
Severity scale, for risk analysis, 190t
SHPO. *see* State Historic Preservation Office (SHPO)
SI. *see* Site investigation (SI)
Sierra Club, 4
Sierra Nevada Forest Protection Campaign v. Weingardt (E.D. Cal. 2005), 246
Significance. *see also* Assessing significance; Significance factors
concept of, 52–53, 204
of cumulative impacts, 192
determining, 42f, 45, 65–66

Index

evidence of, 196–197, 197t
issue of, 50
in practice, 196
Significance factors, 54, 194
   in assessing intensity, 54, 55t
   receptor test assessment and, 182
   regulatory compliance test and, 183
   risk/uncertainty test and, 189
Significant assessment test (SAP), 179–180, 180t
Significant effect, 178–179
Significant impact, 179
Significant issue, 179
Significantly, meaning of, 51
The Silent Spring (Carson), 6
SIP. *see* State Implementation Plan (SIP)
Site alternatives, 86
Site investigation (SI), 255
Site-specific action, 53, 196
Site-specific analyses, 168
Sliding-scale approach, 38, 83–84, 108, 108t, 178
Slip laws, 220
Socioeconomic impacts, corresponding context of, 54t
Soil Conservation Service, 5
Soil impacts, corresponding context of, 54t
Specialized non-NEPA environmental assessment documents
   biological assessment, 262–268
   cell tower NEPA review, 259–262
   environmental assessments and state NEPA programs, 266271
   introduction to, 253
   phase I environmental site assessment, 253–259
Species names, EA documents and, 147–148
Staffing, 137–138
"Standard for Conducting Phase I Environmental Site Assessments", 256
Standing, criteria for, 228–229
State agencies, 139
State Coastal Management Programs, 118
State Historic Preservation Office (SHPO), 135, 260
State Implementation Plan (SIP), 169
Statistical significance, 178
Statutory requirements, 117–118
Strycker's Bay, 226
Style, EA documents and, 142
Subsistence consumption, 125, 126t

Substantive *vs.* procedural process, 225–226
"Sufficiency test", 161
Superfund, 20, 54, 141, 253, 254
Supplemental environmental impact statement (S-EIS), 237, 240
Supreme court decisions, 238–239
Systematic planning, 34

T

Tables, EA documents and, 144–145
Technical information sources, 152, 153t
Technology alternatives, 86
The Fifth Circuit Court Decision, 102–103
Threshold tests, 183
Tiering, 76
Timing, environmental assessment and, 66
Title I, declaration of the National Environmental Policy Act, 17
Title II, Council on Environmental Quality, 17
TOMAC v. Norton (D.D.C.), 246, 247
Trivial violations, 228
Twentieth century, early, environmental movement in, 5

U

U.S. Administrative Procedures Act (APA), 52, 117, 188, 207
U.S. Air Force, 163, 182
U.S. Army Corps of Engineers, 52, 54, 105, 114, 129, 168, 171, 214, 234
U.S. Atomic Energy Commission (AEC), 6, 10, 11, 52
U.S.C. *see* United States Code (U.S.C.)
U.S. Code (USC), 15
U.S. Court of Appeals, 199
U.S. Department of Defense, 189
U.S. Department of Energy (DOE), 36, 52, 76, 114, 115, 189, 190
U.S. Department of Energy (DOE) guidance, 121
U.S. Department of Housing and Urban Development (HUD), 198
U.S. Environmental Protection Agency (EPA), 219, 254
   analysts and, 151
   Congress establishment of, 9
   definition of environmental justice, 119
   guidance, 121
   preparation of EAs and EISs, 52

U.S. Fish & Wildlife Service (FWS), 102, 129, 135, 262, 263
U.S. Forest Service, 63, 64, 100, 231, 245, 247
U.S. Nuclear Regulatory Commission (NRC), 52, 69
U.S. Soil Survey, 4
Unavailable information, dealing with, 37, 38–39, 38t, 112
Uncertainty, 141
United States, environmental movement in, 4
United States Code (U.S.C.), 221, 222
United States Statutes at Large, 220–221
The United States Statutes at Large (Federal Register), 15

V

Venue, 230
Vermont Yankee case, 226
Violations, trivial, 228
Visual and graphic aids, EA documents and, 143–144, 145f
Visual impacts, corresponding context of, 54t

W

Waiting period, FONSI and, 77, 78t
Water Pollution Control Act, 221
Web pages, 140
Wetland mitigation, 214
Wetlands impacts, 54t, 168
World Bank, 15
World Wide Web, 140
Writing an environmental assessment (EA)
  analysis and methodology, 151–157, 152t, 153t
  contents of EA, 148–150, 149t
  introduction to, 133–134
  preparing EA, 134–147
Writ of mandamus, 244

Y

Yellowstone National Park, 4

Z

Zone of interest, 229

Made in the USA
Columbia, SC
30 June 2021